LANDMARK COLLECTOR'S LIBRARY

Victorian Slate Mining

A social and economic study

Ivor Wynne Jones

LANDMARK COLLECTOR'S LIBRARY

VICTORIAN
SLATE MINING

Ivor Wynne Jones

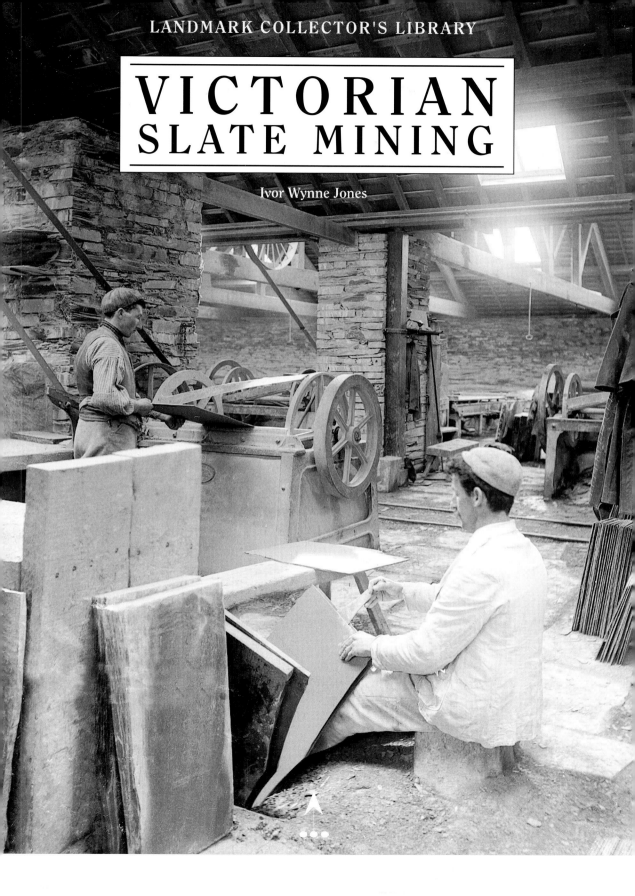

Published by

LANDMARK
Publishing Ltd ● ● ●

Ashbourne Hall, Cokayne Ave
Ashbourne, Derbyshire DE6 1EJ England
Tel: (01335) 347349 Fax: (01335) 347303
e-mail: landmark@clara.net
web site: www.landmarkpublishing.co.uk

1st edition

ISBN 1 84306 073 6

British Library Cataloguing in Publication Data: a catalogue
record for this book is available from the British Library.

Printed by Gutenburg Press Ltd, Malta

Design & reproduction by James Allsopp

Cover captions:
Front cover (and Page 5): An underground chamber at Llechwedd Slate Mines,
photographed in 1894 by J.C.Burrow.
Back cover Top: An 1894 underground bridge to the "safe design" of C.Warren Roberts.
Back cover Bottom: J.W.Greaves's busy quayside at Porthmadog, photographed by J.C.Burrow.
Pages 1 & 2: The magnesium strip lamps used by J.C.Burrow and C.W.Roberts at Llechwedd.
Page 3: Slate splitting and dressing at Llechwedd in1894, photographed by C.W.Roberts.

CONTENTS

ACKNOWLEDGEMENTS

The author acknowledges, with thanks, the valuable encouragement and assistance received from the late Somerville Travers Alexander (Sandy) Livingstone-Learmonth (managing director of J.W.Greaves & Sons), the late John Williams-Ellis (chairman of J.W.Greaves & Sons, and great-grandson of J.W.Greaves), Susannah Brooke (great-great-granddaughter of J.W.Greaves), and his fellow directors Sir Osmond Williams (great-grandson of J.W.Greaves), Aled Ellis and R.Hefin Davies (chairman of the Greaves Group).

Victorian street scenes at Blaenau Ffestiniog. Most of the town centre dates from around 1870. The town grew in distinct communities, each well equipped with rival places of worship, but by 1880 Dolgarregddu had emerged as a natural town centre, complete with two banks. On 24 June 1881 the Post Office formally recognised "Blaenau Ffestiniog" as a postmark, to replace what they had called "Four crosses" since first opening their office in 1866.

With a few minor exceptions, slate mining – as distinct from open quarrying – was confined to the mountains of Merioneth, in Victorian Britain. Its perception as a particularly hazardous working environment led to an immensely detailed public inquiry during 1893-94, conducted by the *Departmental Committee upon the Slate Mines of Merionethshire,* appointed by Home Secretary Herbert H.Asquith, QC.

In November 1893 he announced the appointment, as chairman, of Professor [later Sir] Clement Le Neve Foster, DSc, FRS, of Llandudno, who had been Inspector of Mines for North Wales since 1880 (remaining in office until 1901). Before that he had spent eight years as Inspector of Mines for Cornwall. Four others were appointed to make up the Departmental Committee: John Ernest Greaves, JP, Lord Lieutenant of Caernarvonshire, and son of the founder of Llechwedd Slate Mines, at Blaenau Ffestiniog, which he had himself managed since 1870; John J.Evans, FGS, manager since 1874 of Penrhyn Quarry, in the Ogwen Valley, and before that manager of Dorothea Quarry, in the Nantlle Valley; E.Parry Jones, JP and member of the Local Board; and John Jenkins, who was to become a member of the first Urban District Council in 1895.

Blaenau Ffestiniog schoolmaster Griffith John Williams was appointed interpreter in this community of 11,073 (twice the current population) whose natural mother tongue was Welsh – the only language known to many of the population. Born in 1854, at Bethel Chapel House, Tanygrisiau, but soon moving to Rhiwbryfdir Chapel House, he had come to the notice of the Departmental Committee as author of *Hanes Plwyf Ffestiniog* (i.e. History of the Parish of Ffestiniog), published in 1882, but which he had originally written for a competition at the second Oakeley Quarry Eisteddfod, in 1880. Williams had taken the title and contents – albeit for his bigger book – of a pioneering 1879 volume by Ffestinfab, the bardic *nom de plume* of William Jones, steward or sub-manager at the rival Llechwedd mines, across the road to Oakeley.

Charged with inquiring into "the dangers to health, life and limb of the workers employed in the Merionethshire slate mines," the Departmental Committee held its initial meetings at the Cocoa Rooms, in Blaenau Ffestiniog's High Street. That was the official name for what was always known locally as *Tŷ Dirwest*, meaning Temperance House, an establishment founded by the owners of Llechwedd and Oakeley quarries in this community of thirty-seven churches and chapels, to offer their men an alternative to the town's twenty-two taverns. (Ironically, it is now the Royal Welch Fusiliers Association club). The committee's sessions were held in the upstairs room, which served as the town's magistrates' court from 1892 to 1900. Evidence was taken from three doctors, ten employers or managers, 36 workmen, and three cookery teachers. The committee also held one session at Corris (South of Blaenau Ffestiniog), and visited two slate mines in France (at Fumay and Rimogne) and a marble quarry in Belgium (at Philippeville), to obtain comparative information.

We learn from the unpublished diaries of J.E.Greaves that he, his wife and Evan Thomas, one of his under-managers, had visited the slate quarries at Fumay, north of Mézières-Charleville, on 19 April 1883, when they were entertained by managing director M.Brassart, who also greeted the Departmental Committee eleven years later.

In order to illustrate some of the evidence and the subsequent report to Parliament, Dr. Foster commissioned J.C.Burrow, who had earned fame in Cornwall for his pioneering underground photography in the tin and copper mines, while Foster was H.M.Inspector for the county. Foster no doubt had a great personal interest in photography, for his barrister father Peter Le Neve Foster was one of the principal founders, in 1853, of the [Royal] Photographic Society, during his 26 years as secretary of the Society of Arts. Always an innovator able to capitalise on any opportunity, J.E.Greaves used Burrow's presence in Blaenau Ffestiniog to commission a set of "magic lantern" glass slides of the Llechwedd operation, and these pictures now provide many of the illustrations in this volume.

"As quarrying slate by the underground method is an industry of comparatively recent date in Merionethshire, and as very little is known about it outside the Principality [of Wales], we have thought it desirable to enter somewhat fully into details; for otherwise even an experienced miner would scarcely be able to follow our remarks upon the causes of some of the most common accidents, and our recommendations concerning the best means of preventing them," said the Departmental Committee in their 1895 report to Parliament.

The slate mines of Blaenau Ffestiniog were actually enormous underground quarries – ultimately sixteen floors of them, one above the other at Llechwedd. Although the notion of working stone underground seemed novel in Victorian Britain, the Jews had done it in Biblical days, to obtain the stone with which to build Solomon's Temple, in Jerusalem. Solomon's quarry is mentioned twice in *The Bible*. In the First Book of Kings, Chapter 6, we read: "It was in the 480th year after the Israelites had come out of Egypt, in the 4th year of Solomon's reign over Israel, ... that he began to build the house of the Lord. The house was 60 cubits (110 ft) long by 20 cubits (36½ ft) broad, and 30 cubits (55 ft) high... And when it was in building, was built of stone made ready before it was brought thither: so that there was neither hammer nor axe nor any tool of iron heard in the house while it was in building." That underground quarry of about 2,770 years ago still exists. It was rediscovered in the 1920s when a dog disappeared down a hole outside the city wall, near the Damascus Gate. Widening the hole to seek out his dog, the owner found a labyrinth of impressive workings stretching for some 650 feet in length beneath the city. The entrance is now closed with an iron gate, but access can be arranged.

In the Book of Job, Chapter 28, we read: "The miner searcheth out the stones of thick darkness and of the shadow of death. He breaketh open a shaft afar from where the men sojourn; they are forgotten of the foot that passeth by; they hang afar from men, they swing to and fro. That path no bird of prey knoweth. Neither hath the falcon's eye seen it. He putteth forth his hand upon the flinty rock; he overturneth the mountains by the roots. He cutteth out channels among the rocks and bindeth the streams that they weep not. And the thing that is hid he bringeth forth to light." That could be a reasonable description of the way in which 25 miles of underground quarries were worked at Llechwedd, part of which tourists have been able to explore since 1972 by means of two specially made trains. Victorian mine owners had argued they were operating underground quarries, and therefore exempt from the regime of the Inspector of Mines under the provisions of the Metalliferous Mines Regulation Act, 1872. They lost the argument at the Court of Queen's Bench in 1875, when the workings were declared to be mines, and from which time statistics of fatal accidents became available.

In 1875 there were 4,027 men employed in the Merioneth slate mines, and there were 13 fatal accidents. The blackest year was 1877, when 18 were killed. By 1893, when the Home Secretary decided it was time for a public inquiry, the work force was 4,321 and there had been 163 fatal accidents over the course of 19 years. Following publication of the Departmental Committee's report, the mine owners formed the Ffestiniog District Slate Quarry Proprietors Association, with eleven members: Oakeley (1,591 workers), Llechwedd (500), Foty & Bowydd (488), Maenofferen (361), Glanyrafon (410), New Welsh Slate (292), Craigddu (220), Rhosydd (200), Diphwys Casson (110), Rhiwbach (28) and Wrysgan (102). Another four joined later, and the total labour force peaked at 4,733 in 1898. (Llechwedd peaked at 639 in 1904).

Whether or not the Home Secretary was prompted to launch the public inquiry by the events surrounding the Llechwedd strike of 1893 we shall never know, but the juxtaposition seemed more than coincidental. The 16-week strike ended on 5 September, and two months later Asquith launched his unprecedented investigation. Even while the men were out the *Caernarvon & Denbigh Herald* reported (on 2 June): "The real cause of the strike has almost been lost sight of in the labyrinth of side issues."

The North Wales Quarrymen's Union was formed in April 1874, and three years later had a membership of 8,190, of whom only 378 were from Blaenau Ffestiniog. Rival slate quarry owners competed in the magnificence of their mansions, the most flamboyant of all being the Douglas-Pennant family, owners of Penrhyn Quarry, at Bethesda, and sugar plantations in Jamaica. They built their enormous Penrhyn Castle out of the profits of certain slave labour in one enterprise and what some interpret as near slave labour in the other. However, there was another side to the owners, epitomised in an illuminated address of 1866 presented to the first Baron Penrhyn, from "we the quarrymen and other residents in the vicinity of Penrhyn Quarry," congratulating him on his elevation to the peerage, and adding: "We cannot look around without beholding durable evidences of your munificence and judicious liberality in the shape of churches, schools, hospitals, model cottages and numerous other improvements." Similar benevolent paternalism was practised by other quarry owners, notably by the family of John Whitehead Greaves, the founder (in 1846) of Llechwedd Slate Mines.

A month after the formation of the North Wales Quarrymen's Union all the owners of quarries in Caernarvonshire and Merioneth met at the Royal Hotel, Caernarfon, where each, in turn, pledged to refuse employment to anyone suspected of belonging to the Union. The first test came at Glynrhonwy, Llanberis, where J. W. Greaves had relinquished his lease in 1873. The new owner locked out 120 men who refused to abandon their Union. On the opposite side of the lake Thomas Assheton-Smith, of Y Faenol, owner of the enormous Dinorwic quarries, locked out all his 2,200 workers. On learning that some of the men had found temporary harvest work in the fields of the vast Faenol Estate (which included the summit of Snowdon) Assheton-Smith ordered his tenant farmers to dismiss them. Eventually the Glynrhonwy men were taken back, as Union members, and a compromise was reached with the Dinorwic men, involving replacement of three Union officials and amendments to the Union rules.

A dispute over a Dinorwic quarry rule – prohibiting Union meetings on site – triggered the industry's first prolonged lockout, which lasted from October 1885 to 1 March 1886. The underlying cause was Assheton-Smith's blatant favouritism for those 200 or so men who had not joined the Union, especially if they were Church of England worshippers. Refusal of a pay rise at Penrhyn prompted a series of meetings in 1892, although the real grievances centred upon other issues, such as the 1890 abolition of the traditional monthly holiday, and rejection of an 1891 demand for greater freedom in selecting partners to work the monthly contracts – which were a crude version of four-man co-operatives controlled by management.

The first Welsh Labour Day rally (*Gwyl llafur*), held at Caernarfon in 1892, was virtually a North Wales Quarrymen's Union event, and proved a catalyst in hardening the demands of the men. Five weeks after the May Day rally a deputation of Llechwedd men met John Ernest Greaves, the senior of the founder's sons, demanding a wage of five shillings (25p) a day, which he refused. Later that day the men of the various Blaenau Ffestiniog mines voted by 1,526 to 248 in favour of a strike to enforce their demands. On 25 July the owners met at Porthmadog to coordinate their response and, to the dismay of their Caernarvonshire colleagues, the Blaenau Ffestiniog men settled in November 1892 for an additional 2s a week, taking their pay to 27s (£1.35) for a 6-day week, which was 6d (2^{1}/$_{2}$p) a day less than the goal of 5s. By 19 January 1893 the Llechwedd men were again chasing the elusive extra 6d, in what was described in J.E.Greaves's diary as "a long conflab," followed by a similar demand on 24 April.

May Day of 1893 brought both the 19th annual general meeting of the Union and the second Welsh Labour Day rally to Blaenau Ffestiniog. The AGM was held in the Assembly Rooms, on the upper floor of the now empty 1864 Market Hall, beside St.David's parish church, in what was then called Market Square (but now called Commercial Square, because of the Commercial Inn). In the afternoon the men surged over the bridge for a mass rally in a badly leaking marquee set up on the mud of the recreation ground in New Market Square (now The Square). Three prominent political speakers had been promised by the organisers: Tom Ellis, Liberal MP for Merioneth; David Lloyd George, Liberal MP for Caernarfon Boroughs; and Keir Hardie, Socialist agitator and Independent Labour MP for South West-Ham. None turned up, the two Liberals apologising for other commitments (perhaps to distance themselves from Hardie's class war), and Hardie's sending a message to say he was delayed at Caernarfon. Despite the rain 6,000 quarrymen, accompanied by bands from Bethesda, Nantlle, Llanberis and Waenfawr, marched through the town to the packed tent, where David Griffith Williams, chairman of the Union's Ffestiniog lodge, who had worked at Llechwedd for 40 years, was re-elected Union vice-president.

The men reaffirmed their demand for a minimum wage of 5s a day for quarrymen, with other quarry wages linked on a proportional basis. They also adopted a resolution recognising "the importance of establishing a fund to meet the exigencies of old age and so obviate having recourse to charity." Keir Hardie turned up in the evening and proceeded to incite his beloved working class to work out their own salvation, ridding themselves of oppression by their masters and fighting not only for 5s a day but for more effective control over the mines and quarries in which they laboured.

Two weeks later (17 May) Griffith Jones was spotted leaving early from Llechwedd and was challenged by the steward, William Jones, who told him that if he did not return to his place of work he would lose all his pay for that day. Griffith Jones left and when he turned up next morning was told to go home. He spoke to a growing assembly of his fellow rockmen on their way to the mines, and they were asked by William Jones why they had not proceeded to work, and told they could not hold a meeting during working hours and on company property. By the time mine manager Charles Warren Roberts arrived the men were congregating from many parts of the mine. They were told either to proceed to work or go home – and all walked out for a meeting in Jerusalem chapel schoolroom. A deputation of five, led by Union vice-president D.G.Williams, was sent to Llechwedd to talk to Greaves, who told them he would not discuss a matter still to be resolved between management and Griffith Jones, adding that by walking out the men had all dismissed themselves from his employ. He told them to remove their tools, after which they would be allowed back only to apply to the manager for a job.

Of the 486 men locked out from Llechwedd only 125 belonged to the Union and of these only 75 were paid-up members. The diaries of J.E.Greaves reveal that two days after the men's leaders had called for a fight to victory, the first of the waverers approached him and said they wished to return to work. On 13 June D.G.Williams reported to the strikers that only one man, R.D.Hughes, of Dolwyddelan, had gone back to work, and on 19 June the strikers listed five prominent men from the community whom they invited to speak to J.E.Greaves on their behalf. All refused to intercede; they included E.Parry Jones who, a month later was sitting as a magistrate on cases arising out of the strike, and who in December, became one of the Departmental Committee investigating the working conditions in the Merioneth mines. On 3 July a motion to continue the strike was withdrawn and the men decided to apply for their jobs, but to resume work only so long as no one was refused.

About 24 had passed through the management's individual interrogation before Warren Roberts and J.E.Greaves discovered the conditional nature of the proposed return. Greaves said he had been deceived, adding that D.G.Williams, Ellis Hughes (a local councillor), Rowland Edwards and Robert Pugh would be refused work. That resulted in an immediate stalemate, and a few minutes later Greaves and Roberts saw between 100 and 150 men rushing towards the Floor 2 Mill (now part of the tourist complex), where the only two strikebreakers, R.D.Hughes and Robert O.Hughes were assaulted and threatened with murder, one actually dragged to the edge of a tip. The incident split the men, and although the majority voted to continue the strike, a substantial number met near the office and resolved to go back to work – for which they were given a police escort on site. Violence having tainted the strike, there was a self-defeating incident on 13 July when, in the middle of the night, a gang attacked the Llechwedd Hospital, in New Market Square, smashing the windows. The facility had been provided by the Greaves family, who paid the salaries of the master and matron and provided all medicines and medical appliances for the benefit of mine staff. That night one patient and the resident staff were in bed at the hospital, when a note was pushed under the door stating: "The windows tonight. Fear for worse." Three men were later arrested: Llechwedd striker William Owen; solicitor's clerk J.R.Cadwalader, from one of the community's longest-established families; and Edward Jones, who worked at another of the town's mines. They pleaded guilty at Penrhyndeudraeth magistrates' court to causing damage and were fined £1 each, and ordered to pay for repairing the windows. At the same court seven men were charged with conspiracy to intimidate fellow workmen, in the incident at Llechwedd. They were committed for trial at Merioneth Quarter Sessions in October – where they were defended by David Lloyd George, MP. For different reasons none of the men was convicted, but all lost their jobs at Llechwedd.

Two months later D.G.Williams and fellow accused John Hughes, who had worked at Llechwedd for 27 years, sought revenge by offering themselves as witnesses to the Departmental Committee's public inquiry.

THE PHOTOGRAPHY OF
J.C.BURROW

2

Remarkably little is known about John Charles Burrow, the pioneer of underground photography, many of whose studies appear in this volume. He spent his entire life in Cornwall, where he was born in 1850 and died in 1914. He ran his own printing works at Camborne, where he became popular for his portrait photography. He also used his camera to record disasters, especially the maritime dramas off the Cornish coast. One of his most famous pictures was of the ship *Escurial*, wrecked at Hayle with sailors clinging to the rigging while the lifeboat was stranded on soft sand nearby.

His wider fame stemmed from his underground work in the Cornish tin and copper mines which led to his publication, in 1893, of *Mongst Mines and Miners*, depicting underground scenes at Dolcoath, Cook's Kitchen, East Pool and Blue Hills. It was this work which resulted in his being invited to illustrate the report of the *Departmental Committee upon the Slate Mines of Merionethshire*, at the end of that year.

We can but turn to the brief introductory words in *Mongst Mines and Miners* to discover something about his equipment and techniques: *"Advancement in any particular branch of science is often slow and laborious, and, if examined in detail, the progress is not very apparent,"* he wrote.

"When, however, the distant Past is contrasted with the Present, it is evident that great strides have been made and important revolutions effected. This applies as much to the photographic as to any other branch of applied science; for, although the results of daily researches and experiments may not be very striking in themselves, yet, were our fathers in photography, of forty or fifty years ago, to revisit the scenes of their labours, they would be more than astonished at the present high standard of pictorial work. One is apt to be satisfied with the attainment of any special object, regarding it as the acme of human effort. But there is no standing still, and what seems perfect today will be superseded tomorrow.

"To the scientist, the engineer, and the explorer, the camera has become an absolute necessity. It has been used in all kinds of places, on the surface, under the sea, in the air, and even under the earth's surface. The greater the difficulties encountered, the stronger the determination has been to accomplish successful results. Usually the indispensable light comes direct from the Sun, but that is of course unable to penetrate the rocky crust of mother earth. Hence the first difficulty in underground photography is to satisfactorily illuminate the workings by artificial means. The objects of the following sketch are to describe how this illumination may be effected, and to illustrate the advantages which the artist may take of the illumination.

"After preliminary trials with cameras varying in size from 10 x 8 to half-plate, it was finally determined to use the latter because of its lightness, portability and moderate size, the latter being specially important in confined situations. The light bellows Kinnear form of camera was most suitable. Double dark slides were used carrying fourteen plates, but such were the difficulties experienced that often five or six plates only could be exposed on the same day. When changing plates underground there was no danger of their being damaged by

Left: Photographed by J.C.Burrow, in the Oakeley mines, this study shows an 86-feet long ladder assembled underground to facilitate roof inspections – usually by candlelight.

the light, for when the candles were extinguished there was absolute darkness. But, since it was impossible to keep one's hands clean underground, it was found better to take the plates well protected in dark slides. Lenses of all descriptions were tried, with more or less success. The rapid symmetrical did very well for certain subjects, and a half-plate portrait combination was fairly good when used in an open place where the subject did not require great depth of focus. Care was, of course, taken to prevent the flare spot. The ordinary wide angle was much too slow. A lens having all the advantages of each of these was required, one that would embrace a wide angle, give depth of focus, and speed. This was found in the Zeiss' Anastigmat, Series III, made by Messrs. Ross & Co. It proved to be a perfect gem, and with it splendid results have been obtained, both underground and at surface. A sliding tripod stand was most convenient. Sometimes the camera had to be tilted at an angle of 60^0, and the front leg tied to a rock to prevent overbalancing. At other times it was strapped to a ladder, or a bar, looking down an inclined shaft, or fixed on the ends of a couple of stout planks over a yawning 'gunnies'. The sliding form of tripod allowed the greatest freedom in such situations.

"The best apparatus manufactured is of no practical value without good plates. On this subject, so important to the photographer, pages might be filled with the experiences of a twelvemonth underground. The plates of several makers were tried, slow plates, rapid plates and isochromatic plates. But none equalled the Cadett lightning plate. At first all the candles were extinguished, as the halo round each spoiled the effect of the picture, but, with the lightning plate, every miner was taken in his working position, with his light in its usual place. So sensitive were these plates that the camera had to be well covered to prevent any stray pencil of light being admitted other than through the lens. This requires careful attention when working in daylight. The diaphragm slit in the lens, even, should be covered, say, by means of a wide India-rubber band.

"It is needless to detail the incidents attending the transport of apparatus from surface to the bottom of an inclined shaft half-a-mile below. Some were unpleasant, many were amusing. Occasionally the band of willing helpers was recruited from certain classes of individuals totally unaccustomed to mining and mines. Readers intending to attempt underground photography must not be discouraged when a too eager assistant, staggering through the semi-darkness with a hydrogen cylinder of 20 feet capacity under his arm, and a pair of limelight burners in one hand, stumbles headlong into a pool of water by the side of the level. Nor must they lose their self-control when, after carefully selecting the driest spot available to open the magnesium powder tin for re-charging the lamps, a 'flosh' of water from some unknown source falls into the tin as soon as the cover is off, instantly "dropping the curtain" most effectually upon the remainder of the day's programme .....

"The preparations required were more extensive than one imagines. Each helper undertook a particular duty. One adjusted the limelight burners, another attended to the oxygen and hydrogen cylinders and trappings, another prepared the magnesium lamps, and so on. The writer, after repeated unsatisfactory experiments with different flash lamps, designed a pair of triple-flash lamps, which have proved exceedingly useful and have given most satisfactory results. The high temperatures of the deep mines caused camera and lens to be covered with condensed vapour for some time after the scene of operations was reached, a source of much bother.

"A state of readiness having been arrived at, the word to "light up" produced a powerful flash from the lamps and ribbons simultaneously. The limelights were previously at their maximum intensity. An exposure of from two to four seconds generally gave the best results. If everything appeared favourable in the strong light, another subject was sought; for a second exposure, in the same place on the same day, rarely gave good results, owing to the "fog"

Demonstrating the various chisels used in Victorian slate mining, including the big vertical "jumper", centre left.

caused by the products of the combustion of the magnesium. For close subjects two triple-flash magnesium lamps were generally used, so placed as to destroy shadow as much as possible, but in large areas, such as that shown in Dolcoath Man Engine Shaft, more lamps were brought into requisition. In such places the air was generally cooler and clearer, and sometimes two exposures in succession gave fair results under such circumstances. The bottom of the shaft in Cook's Kitchen Mine was a difficult subject. The temperature there was 100°F. The miners work nearly naked. The camera was attached to the ladder and tilted at an angle of 45°. Water dropped everywhere and came from the footwall in a steady stream. Heat, water, and vapour, combined with the peculiar setting of the camera, made the work tedious and difficult.

"In the foregoing the writer has attempted to sketch an outline which may guide one desirous of commencing experiments in this new channel for photographic enterprise. Very little has been done in this direction so far. The Freiberg and Clausthal mines have each produced a series of plates of close subjects, and Mr. Herbert W. Hughes, of Dudley, has obtained some excellent results in the coal and limestone districts of South Staffordshire. Apart from these and the productions of Mr. Arthur Sopwith, also in coal mines, attempts at underground photography have not generally been successful. So many difficulties have presented themselves at the outset that the work has invariably been abandoned after brief trial."

Manhandling trucks on the
25 miles of underground
railway at Llechwedd.

Because of the vastness of the underground quarries at Blaenau Ffestiniog, Burrow would have been faced with even greater lighting problems than in the relatively compact workings in Cornwall – problems that often defeat modern photographers. While he would have transported his bottles of oxygen and hydrogen for his limelight equipment, and his tins of magnesium powder for his fused flash-lamps, we know, from the 1895 report to Parliament, that at Llechwedd he used a new device produced and described to the Departmental Committee by mine manager Charles Warren Roberts.

Roberts was actually talking about the problem of examining the roofs of the underground chambers, 50-70 feet, or more, above floor level, so as to obviate accidents from rock falls. "I have the Wells light, and it is a very useful lamp. Some few years ago, however, I obtained a magnesium ribbon lamp, which I have found of very great use indeed. It does not carry out everything one wants, but it is an improvement, and can be used at any moment," he told the committee.

"Who is the maker?" asked Dr. Foster, to which Roberts replied: "I do not know. I purchased it from a man named Moses, in Red Lion Square, London, but he has disappeared since then, and I do not know where he has gone to."

"How long will it burn?" asked John Jenkins. "If a man is careful," replied Roberts, "it will go on for as long as the coil lasts, say for a couple of hours. The coil is easily replaced and they cost about 1s 6d (7$\frac{1}{2}$p) each." He told the chairman the lamp had cost about £2.

Two different specimens of magnesium strip lamp have survived at Llechwedd. Each combines a reflector with a brass casing containing a clockwork motor. They have different ways of storing coils of magnesium strip, whose tip emerges from the centre of the reflector. A switch on the casing handle unwinds the strip through the reflector. One or both was used by J.C.Burrow, and Roberts also used the lamps to produce some of his own photographs for the Departmental Committee

Another witness, Evan Hughes, testified to the inquiry that there was a magnesium strip lamp at Maenofferen mine, where he had worked for 25 years. While describing roof inspections he was asked what type of light he used. "Paraffin on cotton waste. We have a rod about four yards long, and we fix it to the top of that," he replied, saying it illuminated up to 12 or 13 yards away.

Asked by the chairman whether that was as good as a Wells light, Hughes replied that he had never used one. "But have you seen it?" inquired the chairman, to which Hughes replied: "I have seen a lamp at our quarry with something like the electric light."

"Was it a lamp with magnesium ribbon?" – "Yes."

"With a reflector?" He agreed, and the chairman said: "So that would be a Wells light?" although a document later submitted to the inquiry showed the Wells light was illuminated by oil, and was expensive in use.

Jenkins joined the questioning: "How long did that last, the light that you saw?" – "It did not work well then. It lasted for five or ten minutes or so," said Hughes.

Clearly there was a technique in making the best of a magnesium strip lamp but it was a device that enabled Warren Roberts and J.C.Burrow to paint in the light on the underground chamber walls, while their camera lenses were open in what would otherwise be total darkness.

Charles Warren Roberts was a civil engineer who worked abroad from 1872, but was installed at Llechwedd by January 1879. He died in May 1897 and was buried at Llan Ffestiniog. None of his negatives seem to have survived, and Burrow's full-plate 8$\frac{1}{2}$ x 6 inch glass plates, which were once carefully preserved at Camborne, were used to make a greenhouse during World War One.

The traditional derrick used for lifting slate blocks in the mines of Merioneth. Their inefficient winches were the cause of many serious accidents. These photographs were taken in 1894 by J.C.Burrow. The scene on the right shows one of the pipes used for pumping water out of the lower workings.

Busy underground scenes at Llechwedd. There were many underground bridges in Blaenau Ffestiniog. The one shown bottom right was to the "safety" design of C.Warren Roberts, manager at Llechwedd. It incorporated iron rod trusses, was fenced, and was fitted with an 8-inch skirting board to reduce the risk of falling stones.

A manager supervising a working rock face from one of the underground bridges. One man, working by candle light, stands on a platform of two short planks resting on two chisels driven into the rock, 50 feet above a sheer drop, while the man bottom right is suspended only by a chain wound around one thigh. In the scene below men are beginning to open up a new chamber.

Worked out into the open, this is one of the spectacular chambers now experienced by tourists riding the Victorian miners' tramway through the Llechwedd workings. It has long been known as Choughs' Cavern, from the rare birds that have nested there year after year.

Victorians believed a direct blood transfusion from a goat would relieve the symptoms of phthisis, or consumption – identified in 1882 as tuberculosis.

Lung problems were an obvious cause of premature death in Victorian Blaenau Ffestiniog. Everyone knew slate quarrymen worked in a dusty environment and while there was a niggling belief that the two were related there was a stubborn reluctance to accept the correlation. When the Departmental Committee convened for its first session, on 5 December 1893, it called as its first witness Dr. Richard Jones, a Doctor of Medicine (Edinburgh) and holder of the Cambridge Diploma in Public Health, who had practised in the town for 14 years. Interestingly, the chairman's first question related to the health of the men – not the accidental hazards.

Dr. Jones submitted a detailed analysis of the death register for the previous ten years. This showed the average age at death of those working in the Ffestiniog mines and quarries was 54 years, if one excluded accidents, and 52.6 when they were included. Lung problems caused 36.4% of the deaths, a third of those being attributed to pneumonia. The death rate per 1,000 inhabitants, from all lung problems was 5.35, and from pneumonia 1.92. Dr. Jones said the figure of 5.35 for all diseases of the respiratory organs compared with 4.6 for London and 3.76 for all England and Wales. The overall annual death rate in Blaenau Ffestiniog was 15.38 per 1,000 inhabitants.

He also tabled what he described as deaths from phthisis. Now obsolete, this word was of Greek origin via such pan-European medieval terms as Middle English tisik, Old French tisike, Italian tisica. All referred to wasting disease of the lungs. By the 14th century phthisis and consumption were being used synonymously for a commonplace deadly disease – for which sweet wine was recommended as a medicine in 1542. In 1861 Florence Nightingale suggested that consumption was induced by foul air in houses. Not until 1882 did Robert Koch, a physician and pioneer bacteriologist working at the German Health Office, in Berlin, announce to a meeting of the Berlin Physiological Society that he had discovered the tubercle-bacillus responsible for pulmonary consumption or phthisis and scrofula. The word *tuberculosis* was born but took a long time to reach the Blaenau Ffestiniog vocabulary, and was only hinted at in a single passing reference during the 1893-94 public inquiry. In Wales the disease was known as *pla gwyn* (white plague), but historically had been given many other descriptions, including *clefyd wâst* (wasting disease) and *peswch y llifiant* (rasping cough).

In Dr. Jones's table we find that during 1883-92 there were 473 deaths among those employed in the local slate industry. Of these 73 were recorded as having died of phthisis, which he further broke down into 43 slate splitters and dressers, 19 rockmen and miners, 10 labourers and one other. The average age of death from phthisis of slate splitters was 34.6 years, rockmen and miners 44.5, labourers 30.16, and others affected by the disease 47 years. The death rate of splitters was 236 per 1,000 inhabitants. However these figures were clouded by Dr. Jones's statement that the rate for deaths from phthisis was 230 per 1,000 for his Blaenau Ffestiniog grouping of merchants, grocers, drapers, tailors, weavers, gentlemen, professional men, clerks, accountants, insurance agents and travellers, and 220 per 1,000 for masons, joiners, shoemakers, bakers, slaters, butchers, saddlers, plasterers, blacksmiths, sett-makers, nailers, carriers and stone cutters. Farmers did best of all at 70 per 1,000, and agricultural and general labourers dying from phthisis were recorded at 190 per 1,000. Just to confuse matters further, Dr. Jones's analysis showed the phthisis death rate per 1,000 was 213.8 for Blaenau Ffestiniog women aged over 14, compared with 161.7 for men.

Responding to J.E.Greaves, who asked whether the Blaenau Ffestiniog population was less strong than in the surrounding rural area, Dr. Jones said: "There are circumstances always attached to mining districts which must be taken into consideration, such as inter-marriage, infectious diseases – taking consumption as one – which tend to make the population not so strong perhaps as an agricultural labourer."

The doctor agreed with the chairman's suggestion: "As far as these figures go they show that the person employed at the Ffestiniog quarries or mines suffers less from consumption than the average person in the district?" Dr. Jones said the women and children appeared to suffer from consumption more than men, adding: "There are other things that ought to be taken into account. There is the geological formation of this district. We know that the dip of the strata is against the natural fall of drainage, and there is always an accumulation of underground water. Our houses are faulty, in that no provision has been made to prevent the evaporation of this underground water. There is no asphalt or anything of that kind in the basement of the houses, and we have no system of drainage." That, he said, was one cause of the prevalence of lung diseases in Blaenau Ffestiniog.

Eventually the question of slate dust was actually mentioned. Dr. Jones said: "This question of phthisis I refer to particularly in relation to the condition of the men who work underground, compared with those who work in the mills – dust inhaling. To my mind it is a very difficult thing to prove exactly the relation between these two factors – the cause and the effect – because in the first place one is struck by the absence of initiatory symptoms, such as chronic cough, asthma, and so on, which we get in other occupations where phthisis is met with. We get premonitory symptoms which ultimately lead to consumption. In Ffestiniog these are absent in great measure, though not entirely, and one is struck by the fact. I know of only one case of emphysema in the whole of Ffestiniog. I do not know of anyone who has performed a post-mortem examination of the lung of a quarryman in this district, and in the absence of that I find a difficulty in arriving at a proper conclusion."

He noted "the want of proper ventilation in the bedrooms of the quarrymen, the dampness of the surrounding soil, the nature and quality and mode of cooking their food, and the cold and wet they are exposed to. In addition to this there is the question of slate dust. How far that goes as a consideration I cannot prove. Slate dust will undoubtedly irritate. That is the primary effect of it, and perhaps it will set up inflammation which will lead to chronic phthisis, which we find the quarrymen of this district suffering from."

Greaves interjected: "That is entirely presumptive?" to which the doctor could only reply: "I have no means of proving it." He also agreed with Greaves that his statistics showed quarrymen to be remarkably free from these complaints, but the chairman observed that the death rate from consumption was higher among men working in the splitting and dressing sheds, compared with those working underground.

"Is it certain that it is the dust that makes the difference between the two men?" asked the chairman. "I would not like to say so," replied the doctor.

Asked by Greaves whether the district was afflicted by other infections, such as typhoid, Dr. Jones said the Local Government Board appointed a Dr. Bloxall to investigate the matter in 1875, resulting in his reporting that an average of 12.9 people a year had died from typhoid and typhus during the period 1865-74. Dr. Jones's own research showed that between 1880 and 1890 the average dropped to 1.3 a year – after the introduction of a piped water supply from Llyn Morwynion, before which the community relied upon shallow wells and polluted streams [carrying the water home in heavy buckets, and therefore being frugal in its use]. However, when asked by Jenkins if there had been more recent cases, the doctor said there had been eight or nine over the previous few months. Asked for details, he said he knew of a girl of 14 years, boys of three and 12, and men of 35 and 45.

Presumably he knew that ever since 1827 Wales had been talking of *clefyd Stiniog* (Ffestiniog disease), which became the term for typhoid fever. During the summer of 1863 there were more than 600 cases in the parish, resulting in 67 deaths. In 1871 the local eisteddfod held a competition for a paper on how best to improve the health of Blaenau Ffestiniog, but unfortunately the winning entry has been lost. Dr. Jones could also have imparted more information from Dr.Bloxall's 1875 report, in which he said typhoid returned to the area every summer, and there were other diseases, such as 900 cases of measles during 1872-73. One of the main causes, reported Dr. Bloxall, was dirty overcrowded housing, in which disease could spread rapidly to everyone, with six or eight inhabitants suffering simulta-neously in any one house. Dead bodies were kept in the crowded houses for several days, for want of a local mortuary. There were other causes of disease, such as the practice of empty-ing household latrine buckets into streams and ditches, or on to one of the scores of rubbish tips dotted about the town. The backyard latrines were themselves filthy, and were just as bad at the school, said Dr. Bloxall, writing eighteen years before the 1893 public inquiry.

Reverting to Dr. Jones's evidence, he was asked by J.J.Evans: "Do you admit generally that dust has a bad effect on the lungs of the men who have to work in it?" Again he parried the question, stat-ing: "I have said before that I believe the primary effect of the slate dust is a me-chanical effect. It irritates the part it comes in contact with, and it may be the means of setting up inflam-mation, which may develop into other diseases."

"Do you think slate dust is less injurious than gran-ite dust?" he was asked. And answered: "It is cer-tainly not so injurious ac-cording to statistics that have been prepared on the point. I work it out that amongst the Ffestiniog quarrymen, out of 1,000 deaths from all causes, 154 are due to phthisis, whereas the proportion of the Cor-nish miners [between 25-65 years of age] is 375 per 1,000."

"Have you ever had complaints from the men themselves that they consider the dust affects their health, from their point of view?" asked Evans. "Yes, I find the men complain of dust," he replied. "They think the dust is the cause of their illness, and many of them seem to believe that if they take a drink to wash the dust out of their mouths, they improve their condition – say a glass of beer or water. It gets into a state of clay and clings to the roof of the mouth."

Asked if the dust could develop into pneumonia, the doctor said it could not be the cause, but might develop with other causes, such as a chill. Asked if pneumonia was prevalent in Blaenau Ffestiniog, he said 55% of the deaths he noted in 1889 were due to pneumonia, but that was an epidemic year. During the period 1880-89 only 36.4% of local deaths were due to pneumonia. He had no information with which to compare that with the rest of the country.

"In your opinion the dust has nothing to do with it?" asked Jenkins. "I do not say it has nothing to do with it, but I should say it would require other conditions."

Another witness called on the opening day was John Owen, a slatemaker at the Oakeley quarry. "Have you suffered in any way from inhaling dust?" he was asked. "No, I have not," he said.

"Do you, in your experience, know of any men who have suffered from the inhalation of slate dust?" – "I cannot refer to anyone in particular."

"Do you feel the dust to be very agreeable?" – "I do not know of anything that will do away with it."

"Are the men in the mills sometimes to be seen white with dust?" – "No, except when the door is open, and the wind blows the dust on to them."

"I suppose there is a quantity of dust on the coats of the workmen when they leave the sheds; it colours their coats, does it not?" – "Yes, there is a certain quantity of dust from the slate that cannot be avoided. If a man gets his trousers wet, and he beats it after it dries, a good deal of dust will arise from it."

Asked why he had come to give evidence, and whom he was representing, John Owen said he was invited by the Oakeley quarry manager.

"Do you know the general feeling of your fellow workmen about the agreeableness or otherwise of dust to work in," asked Jenkins, to which the witness replied: "No, I do not. That is how we have always been working, and I have not thought of any means of doing away with the dust. There must, of necessity, be some little dust in working slates, but in my opinion there is not so much now as there used to be under the old system."

"Is there, as a matter of fact, a good deal of dust in the mills still?" – "Yes, there is, especially in the summer, and it is possible there is much more than we can detect."

"If Mr. Williams [i.e. the interpreter G.J.Williams] were to go into the mill with black clothes and stay there for some hours, would they be black when he left?" – "They would not be so black as they are now, perhaps."

On the third day of the inquiry, on 20 December, Dr. Robert Roberts was called to give evidence, the chairman's first question being: "Can you tell us what, in your opinion, are the principal dangers to health, life and limb entailed by occupation of the quarrymen of this district?"

Dr. Roberts, who was the resident doctor for the Factories Inspectorate, immediately homed in on the question of slate dust. "The occupation of a Ffestiniog quarryman must be classed among the 'dust inhaling' occupations," he said. "I contend it may be fairly inferred that this inhalation of dust, smoke, and the products of explosives underground, together with the natural dampness and coldness of the slate rock itself, with which he (the quarryman) is in continual contact, must have a deleterious effect on his health."

The finest particles of dust were angular, said Dr. Roberts, and must give rise to irritation and coughing when inhaled deeply into the air passages. "But from my experience, and the testimony of the workmen, even the youths, it does not do so to any appreciable extent, which may be accounted for by its comparative softness and the ease with which it mixes up with the natural secretions of the air passages to form a clayey pulp favourable for expulsion."

He said the irritability of different kinds of dust was governed by the amount of free siliceous matter, and in the Ffestiniog slate he did not think there was any free quartz. Here, at last, was the first reference to "siliceous matter," the cause of what we now know as the crippling lung disease silicosis – a term used elsewhere at least as early as 1870. The good Dr. Roberts was not a geologist and he did not have the benefit of a modern chemical analysis of Blaenau Ffestiniog slate. A 1922 analysis tells us it comprises 55.3% silica. (The other parts are 24.84% alumina, 10% iron oxide, 2.46% magnesia, 1.47% potash, with traces of soda, lime, sulphuric acid, and 4.7% water). It is the high percentage of silica, against which today's splitters protect themselves with masks and water, that gives Blaenau Ffestiniog slate its superbly fine splitting property – probably the world's best roofing slates.

Continuing his evidence, Dr Roberts said many men complained of lassitude and thirst, and it was hard to know why, unless it was the extent to which they had to depart from normal hygienic conditions. "The atmosphere in which the Ffestiniog quarrymen lives during working hours one may say is generally clouded in dust. In 1890 I wrote a paper on *Slate quarrying as a dust-inhaling occupation*. The only figures I could then get were that among an average of 3,500 quarrymen over 14 years of age there had been, during a period of ten years, 102 deaths from phthisis." Of these 53 were slate splitters, 29 miners, and 14 labourers. "I have reason to believe the registrars of the district do not, even to this day, classify the deaths with the required accuracy, because about 50% of the quarriers [splitters, sawyers and dressers] still appear on the register as dying from phthisis as against the miners and labourers combined, and the question remains to be asked – is it a fact?"

The object of his paper had been to compare the death rate from phthisis alone among the different sections of the men employed in the quarries, and not to compare them with those not so employed. Next to indigestion, rheumatism was the most common complaint. "The fitful nature of the quarryman's work renders him peculiarly liable to chills and their consequences which will produce the next numerous and most serious and fatal class of diseases: pneumonia (inflammation of the lungs), bronchitis, pleurisy, and the usual pulmonary infections, including phthisis and its varieties, but I may say I never had the opportunity of making or learning by post-mortem examination that any of the Ffestiniog quarrymen died of a silicosed (or stonemason's) lung, although I entertain strong opinion such must have happened and will happen," he said, in a pioneering suggestion of what we now know to be fact.

Dr. Roberts told the committee: "The slate quarryman, who of necessity must have his whole apparel saturated with the different kinds of dust floating about him, has his skin also covered by the same, and he ought to have a thorough wash every evening and change his clothing, so that he may have his day clothes dried, well aired, and brushed to put on again next morning. From a sanitary point of view he needs it as much as the collier, who enters his tub every evening on returning from work. The quarryman who neglects proper ablutions has a fine layer of clay on his skin, but not so conspicuous in colour, of course, as that of the colliers. Ablutions are sadly neglected, and it is to be deplored that we have no public baths erected in such a populous district, and one which, from the nature of the occupation of its inhabitants, has a special claim to it."

Greaves observed: "You have referred to the dust in the mills as being injurious to the quarrymen, and calculated in some measure to affect their lungs?" – "I stated that I believe it to be deleterious to their general health."

"Can you instance any particular case in which such an effect is positively traceable to that cause?" persevered Greaves, to which the doctor could only reply: "No. I think I have stated that I have never had such a case. It cannot be proved without a post-mortem examination of the lung, and I have never known of a case where that has been done, with that view, in our neighbourhood, but I have a strong suspicion that such cases do happen."

Evans asked if a post-mortem examination of the lung of a quarryman would show it to be affected by slate dust. "Not upon every lung," replied Dr. Roberts, "but there are some cases I have met with in which I should say *'I believe if I could see that man's lung I should find evidence of the effect of dust upon it,'* but I have never had the chance of making an examination of that kind." He added that for the advancement of science he would like to see an end to the prejudice against post-mortem examination.

Jenkins asked: "Is there any doubt whatever about the unhealthiness of all dust-inhaling occupations?" – "No, there is not."

In response to further questions Dr. Roberts said a post-mortem examination was necessary "to prove that a man has died from any particular disease through the effects of dust – say phthisis." Although not articulated as such in today's vocabulary, that was the first publicly recorded suggestion that some of the Blaenau Ffestiniog deaths attributed to phthisis were caused by silicosis, not tuberculosis.

Continuing his questioning, Jenkins asked: "Do you contend it is possible for the dust not to go to the lung if it is not inhaled?" – "Yes, I do, as regards the Ffestiniog slate," replied the doctor. "I have already said the irritating quality of dust depends upon the amount of siliceous particles in the dust."

"Therefore," said Jenkins, "you rather disagree with the quotation you gave in the paper you read some time ago, in which you mentioned that Dr. Aldridge states that even particles of flour will mix with the moisture in the bronchial tubes, and form a sort of clay, and therefore clog instead of injuring the tubes and develop emphysema?"

Refuting the suggestion, Dr. Roberts said: "I refer now particularly to phthisis, but as regards slate dust I have been writing to Dr. Aldridge that I class the slate quarrymen in the same class as the miller. I regard the action of slate dust as resembling that of flour, in the absence of siliceous particles. At the same time I believe there are cases in which slate dust gets into the lung, if we had only the opportunity of proving it."

Asked if there was any danger of the irritation caused by inhalation of slate dust developing into pneumonia, Dr. Roberts said: "Not into pneumonia, as I suppose you are now thinking of. If you are thinking of the disease we call pneumonia, we look upon it as a kind if fever, or blood poisoning. The [slate dust] irritation sets up a kind of chronic inflammation leading to phthisis, and not to pneumonia as a fever.....When the siliceous particles get into the lung they remain there and cause irritation. This sets up a kind of local inflammation to start with, which develops ultimately into what we call the fibroid phthisis, which is peculiarly the state of the lungs of those who die from phthisis produced by the inhalation of dust."

"Could the quarryman preserve himself to a certain extent against the ill-effects of dust by keeping his mouth shut and breathing through the nose?" asked the chairman, to which the doctor replied: "Certainly. The nose, I should think, has been intended by nature for use that way."

Greaves asked whether slate dust was any more injurious than street dust in a large city, to which Dr. Roberts replied: "The quarryman is continually inhaling it." [At that time most streets were made of loose stone].

Dr. Roberts was responsible for examining and certifying boys of 14 who wanted to work in the mines and quarries of Blaenau Ffestiniog. In response to questions he said agricultural labourers coming into the slate industry were generally healthier than boys from the town.

"Have you ever noticed any of those agricultural labourers after they have come to work at the quarries? What would be the effect upon them of working for some time underground or in the mills?" asked Jenkins.

"I think they generally lose their colour in a year or two. We do not meet with very many people in Ffestiniog who are what we call robust or plethoric, men who are full of blood and require bleeding when they are taken ill," replied Dr. Roberts.

Nineteenth century medicine was still somewhat primitive. Dr. Robert Roberts was one of the most respected surgeons and physicians in North Wales. When asked for his qualifications, at the 1893-94 inquiry, he said: "I have been here all my lifetime. I was apprenticed here first of all, and I have been in practice as a qualified man for 31 years." He was apprenticed to his father, Dr. John Roberts, who was described as a "doctor" on Robert Roberts's 1839 birth certificate, but amended to "bone setter" in a later family marriage certificate, after the Medical Act of 1858. This legislation required what is now known as the General Medical Council to assess and register physicians and surgeons, many of whom had practised without any training, and without ever stepping outside their village. Robert Roberts had therefore served his surgeon's apprenticeship with a man who was refused registration in 1858 – when surgeons were, in any event, regarded as inferior to physicians, a hangover from a century earlier when surgery was usually performed by barbers. Dr. Roberts was a direct descendant of Robert Roberts, of Isallt, near Brynkir, the first of the family to trade as a doctor, and described as "surgeon" when he died in 1776.

Our Dr. Robert Roberts of the 1893-94 inquiry qualified as a Licentiate of Midwifery, at the Andersonian University, Glasgow, in 1859 – after an apprenticeship with his mother Betsan, who practised as a surgeon and midwife, during her husband's years at Congl-y-wal. Three years later he was registered as a Member of the Royal College of Surgeons (England). He was 37 years of age, before he was registered as a Licentiate of the Royal College of Physicians (Edinburgh) in 1876. He was one of eleven children, four of whom (Robert, William, John and Griffith) were apprenticed to their father to practise as doctors and surgeons.

Called to the witness table, John Ellis Roberts, of Bryngoleu, Tanygrisiau, was asked if he had ever suffered from the effects of dust. "I cannot say for certain whether the dust has affected me or not, but I believe, judging from the spittle, that a good deal of dust goes where it ought not to go," he said.

"But if you spit the dust out, do you not get rid of it in that way?" asked the chairman, only to be rebuffed: "No, I do not believe that, as a good deal of it goes to a place from where it cannot be spit out."

Jenkins asked whether too much ventilating wind in the mills tended to raise dust, but the witness said the wind removed the dust and prevented the men from inhaling it, although it created the problem of causing colds.

Yet a third doctor was called to give evidence to the Departmental Committee. He was Dr. Robert Davies Evans, MRCP, LRCS, LSA, of High Street, Blaenau Ffestiniog, the son of local draper John Evans (although born at Llanystumdwy, his father's having married five times). R.D.Evans's son Dr. (later Sir) Thomas Carey Evans was married in 1917 to Olwen Lloyd George, daughter of the Prime Minister. He told the committee he had 1,000 patients, nearly all quarrymen and their families.

Dr. R.D.Evans said half his patients' deaths were from lung diseases, with consumption/ phthisis accounting for 24.1% of deaths among males, and pneumonia 12.8%."The men are, generally speaking, strong and free from a tuberculous or scrofulous taint, and, as Dr. Roberts, the Factories Medical Inspector, said yesterday, only strong and healthy boys are certified by him for admission to the quarries.....The men and their families are comfortably housed in houses that they mostly own themselves, and which are model workingmen's houses....The men are of highly moral and temperate habits." [There were actually 2,380 houses in Blaenau Ffestiniog urban district at that time, of which only 520 were owned by the occupier].

Dr. Evans said there was difficulty in obtaining pure water for the men at their places of work, where, on a hot summer's day they could not quench their thirst without risk of taking contaminated water. Owing to Blaenau Ffestiniog's heavy rainfall, of some 80-100 inches a year, the faeces deposited on the surface were washed away "otherwise the accumulation of them during the years would have been immense," but the polluted water percolated through the crevices and some of it ran to the underground workings, where it was probably drunk.

"In the summer time I have noticed we have a great number of men who are attacked with diarrhoea and vomiting – English cholera – which we cannot attribute to the eating of unripe or bad fruit, inasmuch as the men do not eat anything of that kind. They cannot get anything of the kind in the quarries, and they do not buy it in the shops in the village. I attribute it to the men drinking water which was unfit for them."

Dr. Evans said the majority of deaths from consumption were among men working in the sawing, splitting and dressing sheds, while the majority of pneumonia deaths were among men working underground. Asked how he would make the sheds healthier, he said: "Diminish the dust....There is no doubt the men work in dust all day long. They are, as it were, in an atmosphere of dust, and deaths from respiratory diseases are always high when men inhale dust. There is a great difference between slate dust and, say, coal dust. If you look at the dust of stone and the dust of slate under the microscope you will find sharp angular points in both, whereas in the dust of coal you will find rounded corners, which makes the dust very much less irritating to the respiratory passages."

In answer to another question the doctor said: "You have colliers dying from chest diseases as well as the quarrymen, but the proportion among colliers is very small compared with quarrymen. This, I believe, is accounted for by the fact that coal dust is round and soft, and consequently not so irritating to the mucous membrane of the air passages."

Today's rapidly declining population of crippled pneumoconiosis pensioners in the defunct South Wales coal field would see Dr. Evans's differentiation as Victorian scientific naiveté, for any deposits in the lungs, including those left by tobacco smoke, are obviously harmful to the organ's intended function and therefore to general health.

Asked by Greaves for evidence to support his assertion that slate dust was harmful, Dr. Evans said he could not do so until such time as post-mortem examinations were allowed. It was his strong opinion that slate dust did affect the health of quarrymen. The death rate among quarrymen was double that of others who inhaled atmospheric dust such as wood-dust, wool-dust, or cotton-dust. Dr. Evans said the miners were more subject to emphysema than the rest of the quarrymen. "I saw two last Sunday, and both had marked emphysema. I made enquiries as to the place they worked in, and found they had worked for 30 or 40 years in the levels."

Robert Roberts, manager of Oakeley quarry, was reminded by the committee that they had heard a lot about dust, and asked if he could suggest any means by which it could be lessened "or rendered less of a nuisance." He replied: "You might run a pipe along the length of the shed and cause a spray to play on the dust, but personally I would rather have the dust than work in a damp place."

David Griffith Williams, the union leader sacked a few weeks earlier at the end of the Llechwedd strike, said there was no proper ventilation in the sawing and splitting sheds. "The only ventilation they have is through the doors, and in no other way. If the doors are closed there is no ventilation at all, and the place is too close."

"Is there enough wind?" asked Jenkins, to which Williams replied: "When the door is open there is too much."

"What would you call that, draught or ventilation?" – "Draught."

"With regard to the laying of dust it was suggested here yesterday that a door should be made at the back, behind the dressing machine; would you recommend that?" – "Yes, I would. I think it would be very desirable, because the wind blows the dust up to a man's face when he is working at the dressing machine."

Recalled by the chairman, Dr. Richard Jones said he believed there was a pneumonia bacillus.

"Is it not an opinion of some that there is a bacillus living in the earth that brings on tetanus?" – "Yes."

"Is there any reason to suppose that there is a bacillus underground that produces pneumonia?" – "I do not know."

E.Parry Jones asked about dampness, to which Dr. Jones said that would be one factor in the production of pneumonia "the same exactly as dust or smoke, or the combustion of gunpowder." He blamed the wind for the pneumonia epidemic in June 1889, saying the germ must have been wafted into Blaenau Ffestiniog during 17 days of north, northeast, and east winds, when the atmosphere was very dry and the nights excessively cold.

As recently as 1922 the North Wales Slate Quarry Proprietors' Association, founded in 1911 [and reorganised in 1917], issued a statement in which the Penrhyn Quarry doctor said: "I became convinced, after four years experience here, that slate dust is not merely harmless but beneficial."

Writing for the National Museum of Wales, in 1925, Dr. F.J.North, the Keeper of Geology, said: "Slate powder is capable of destroying the active properties of toxins, such as that of diphtheria. To this may be attributed the noticeable rarity of sickness and septic wounds amongst the workers in the slate quarries." However these words were deleted from the second edition of *The Slates of Wales* in 1927, when the Museum Director, Cyril Fox, stated in his preface: "Fresh light has been thrown upon some of the problems concerning the slates of Wales." It was December 1939 before the word "silicosis" first appeared in the minutes of J.W.Greaves & Sons, the owners of Llechwedd Slate Mines – in response to new legislation requiring quarries to provide relevant insurance cover for their staff against what was, at last, scheduled as an industrial disease.

By that time slate dust was actually being manufactured – known as fullersite – for various industrial uses. In 1941 it was recommended as a much lighter substitute than sand (an important factor in a roof space) for extinguishing German 1-kilo magnesium incendiary bombs raining down on British cities.

TEA-DRINKING PERILS 4

Tea, which had been Britain's national beverage since the mid-18th century, was one of the greatest evils to arrive in Victorian Blaenau Ffestiniog, according to witness after witness at the public inquiry. "The common diseases of the quarrymen are diseases of the respiratory organs in which pneumonia plays a very important part; rheumatism, with all its complications; and diseases of the alimentary system, such as indigestion, and so on," said Dr. Richard Jones.

"I believe a great deal of the latter class of disease is due to the way in which the quarrymen prepare their tea. It is the habit of the quarrymen in this district to send a boy, about half an hour before the meal time, to the *caban* (one of several small eating houses dotted about the mines), prepared by the owners for their comfort, with tea and sugar and water in the same kettle, which is put on the fire and boiled. It then stews there half an hour or more before the men come to drink it. That habit, continued day after day for a number of years, is in my opinion the cause of the indigestion we find in this district, and other diseases that come as a consequence. They undoubtedly ought to have two kettles. They ought to boil the water in one and brew it in the other. I have been trying to preach this to them for a long time but they will not be advised on the point," said Dr. Jones.

"They take bread and butter and tea far too often, and this leads to a want of stamina and inability to withstand disease when it comes," added the doctor. Asked by Evans if the quarrymen suffered from nervous debility, Dr. Jones said that was so, and due, in great measure, to their way of brewing tea.

John Owen, a slatemaker at Oakeley quarry, was asked by the chairman if he had ever been under the care of a doctor. "Yes, many times," he replied.

"From what complaint did you suffer?" – "Indigestion."

"And do you know what brought it on?" – "No, I do not. I am very subject to it always."

"What are your meals generally?" – "Tea, mostly."

"What do you take at breakfast?" – "Tea."

"And what do you get with it?" – "A little meat."

"You do get meat?" – "Yes."

"Do you get meat every day of the week?" – "Yes, I do, a little."

"What does your dinner consist of?" –"Tea."

"And what else?" – "Bread and butter."

"Do you get any meat with that meal?" – "Yes, a little meat or a little cheese, just as it happens to be."

"What do you get for dinner after you reach home?" – "Potatoes and meat."

"Do you take tea with it?" – "No."

"Do you take any supper after that?" – "Yes."

"What do you get then?" – "Tea, as a general rule."

Left: One of the many sources for the tea that was believed to be a great health hazard in Victorian Blaenau Ffestiniog.

"Did the doctor tell you that mode of living was likely to bring on indigestion?" – "Yes, he has told me many times that drinking too much tea would bring it on."

"Do you believe it?" – "Yes, I do, but it has become as second nature to me."

"You do not think you could alter your way of living if you tried?" – "Yes, I could."

"And you think you would live longer if you did so?" – "I cannot say anything about that."

Richard Jones, of Capel Garmon, a rockman at Middle Oakeley, agreed with the previous witness's description of the quarryman's mode of living, but said he lived slightly differently.

"Are you fed better or poorer than he is? Do you take beer instead of tea?" – "No, I take buttermilk." [Buttermilk was what was left in the churn after butter had solidified out of the cream, in the days when nearly every small farm churned butter from its own milk].

Dr. Robert Roberts said the Ffestiniog quarryman was open to the common reproach of drinking too much tea. "No doubt he deserves it, but probably to no greater extent than his neighbours, and I believe there is too much truth in the charge that the older quarrymen have, by feeding themselves so exclusively on bread and butter and tea, spoiled their appetites for the more substantial kinds of food," he said.

The chairman asked whether a good meal when the men returned home in the evening would be better than a good meal in the quarry at midday. "A cup of tea would do very well at midday provided it is well made?" asked Dr. Foster.

"Yes, it ought to be made properly," said Dr. Roberts. "It should not be infused for more than five minutes. By the present mode of preparing it, it loses its volatile oil and its stimulating property."

"Would the rule as to the five minutes apply if you were to make a kettle of it?" asked the chairman, the doctor's replying: "It would not make any difference at all, no matter how much you make."

"Would it infuse in the same time, no matter how large the quantity?" – "Yes, exactly the same. I believe that tea ought only to be infused for five minutes, and it ought afterwards to be drunk there and then. It is not only a stimulant but a nutrient, and it retards the waste of tissues of the body."

John Ellis Roberts, a slatemaker from Tanygrisiau, was asked how he made the tea he took up to the quarry. "Does it stew very much?" asked the chairman.

"The way the tea is made at the *caban* is one reason why I eat at the other place," he said, having earlier described how he preferred to eat his sandwiches with a few friends in the saw-sharpening room. "I make the tea in the same way as they do at home. We boil the water in the kettle, and let the tea brew for a few minutes in an ordinary earthenware teapot."

"Do you prefer a cup of tea at dinner-time to anything else?" – "Yes, I find it better to work after. I used to take buttermilk with me, but I found that rather heavy and it made me drowsy."

Dr. R.D.Evans was asked: "Do you think the quarryman suffers from drinking bad tea?" and replied: "I believe they drink too much tea, and that the tea they drink is not properly infused, but I do not think myself that half the indigestion is due to tea drinking."

"But some of it is, you think?" – "Yes. A great proportion of it."

Oakeley manager Robert Roberts was asked if he had provided all the dining accommodation the men required. "Yes, that is so. There are many men in the quarries who will not avail themselves of what we have there at the present moment. They prefer to eat apart from the rest of the men. What their reasons are I cannot say. Their tea, and so on, they prefer to warm over a candle underground, rather than come up to the dining room."

Eleanor Russell, a cookery teacher at the local schools, said many of the town's inhabitants seemed to eat meat only on Sundays "and their food all through the week afterwards is tea and bread and butter." Like their fathers in the quarries, the children took a can of tea to school and drank it in their lunch-hour, having just bread and butter with it.

Winifred A.Ellis, a cookery lecturer working for Merioneth County Council, told the inquiry: "Incessant tea-drinking is undoubtedly becoming a real calamity to the physique of men and women. To neglect porridge, oatmeal cake, *bara-llaeth* [bread & milk], *cawl* [broth], and shot [an old nautical term for weak pea soup], in favour of tea three or four times a day is to destroy the stamina, to induce indigestion and dyspepsia, and to bring about enfeeblement of body and mind."

"Tea has such a charm for such people that they are sensitive about putting away the cups and saucers when they are not actually in use," continued Miss Ellis. "Tea often serves as breakfast, dinner and supper, the only accompaniment being bread and butter, and sometimes tinned meat. This is obviously insufficient for a hard-working man or woman, and the harder a man works the more food he will need, and that wholesome and nourishing. I regret that in some cases the women prefer to cook pancakes swimming in butter, with tea, than a good substantial dinner." In answer to a question she said pancakes should be confined to Shrove Tuesday.

After the conclusion of the inquiry, on 3 April 1894, County Councillor Robert Thomas, of Corris, submitted a written statement saying: "Remarks have been made here and at Ffestiniog charging quarrymen with excessive tea drinking. I consider this requires a qualification in justice to the men. They are forced to it by the nature of their occupations. They work under very different conditions to farmers and mechanics. The underground men are working in a dense, foggy, and frequently unwholesome atmosphere, and the engine sheds being dusty, causes thirst, with weak appetite and digestion. A butcher here told me that he cannot sell fat English mutton to quarrymen, their stomachs being so delicate.

"I say it is not fair to attribute to quarrymen the fact of being possessed of an unnatural and excessive craving for tea drinking. Tea allays thirst. It is a necessary stimulant, though not much more, but it helps the assimilation of nourishing food. But it is quite erroneous to suppose that quarrymen and their families in this district, however it may be in other centres, depend on tea, cakes and tinned meat."

The perils of drinking tea in slate mining communities had been articulated as early as 1871, when the Reverend W.R.Ambrose, of Talysarn, in the Nantlle Valley, writing about the tea-addiction of the local quarrymen, said: "In the morning, in their haste, they drink one or two cups of tea, and take tea with them in cans to drink during their meal break. Very often when they return home they have nothing but tea again, and we know some of them who are so fond of the product of the leaf that they do not care about anything else. Too much tea is poison."

In a pamphlet written at Dinorwic Quarry Hospital, Llanberis, in 1894, Dr. R.H.Mills Roberts warned that quarrymen should not drink tea that had been stewed. Five minutes, he said, was long enough for tea to stand. He said they needed more substantial nourishment than bread and butter and tea.

A century later (in May 2002) a research project at Harvard University showed that heart attack patients who drank the most tea were least likely to die in the next three or four years. "Tea is a good source of flavonoids, an antioxidant that prevents blood clotting, and that may stop the hardening of the arteries," said the researchers. Heart patients who drank one or two cups of tea a day had a 28% lower rate of dying than non-drinkers – regardless of age, sex, smoking status, obesity, hypertension, diabetes or previous heart attack.

In 1870 John Whitehead Greaves, founder of Llechwedd Slate Mines, built himself a handsome house at the entrance, well away from the stench of the town. He called it Plas Weunydd and it now serves as the headquarters of the family company bearing his name. The illustration below is a cross section of the mines, before the development of floors 5, 6 and 7. Modern tourists can descend as far as Floor B.

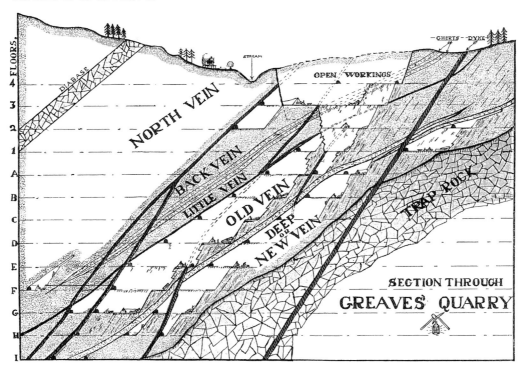

SECTION THROUGH

GREAVES QUARRY

GAY WIVES AND STINKING HUSBANDS

5

In the isolation of Victorian Merioneth the quarryman's wife was a vain creature, according to cookery lecturer Winifred Ellis – although the thought was already in the mind of the Departmental Committee, as evidenced by John Evans's question: "Do you think they dress in the best manner for themselves, or do they go in for fineries?"

Miss Evans replied: "I think they are fond of gay dress, whereas a quieter dress would be more becoming." In those days "gay" meant nothing more than merry, bright and showy, but with a slight hint of loose living. Gay, she may have been, but the Blaenau Ffestiniog wife lived amid an appalling stench, washing her husband's shirt only once a month, or every two weeks if she were fastidious. Furthermore she went to bed with him in the shirt, flannel vest and flannel drawers he had worn all day while sweating in the quarry, his body permanently caked in clay. None of them ever had a bath.

Laundering was a laborious process. In working class Britain, because of the difficulty in obtaining large quantities of hot water, clothes were usually washed in cold water. Although tax on soap was abolished in 1853 it remained the tradition to use it sparingly, and perhaps not at all, using lye instead. This was a homemade alkaline liquid drained from a mixture of wood-ash, obtained from grates or ovens, with water, and perhaps urine, and sometimes domestic fowl droppings. Clothes were placed in a dolly tub, or small barrel, in which they were agitated with a three-legged dolly peg which had a handle at the top – a device still in use in Britain half-a-century ago. If the gay clothes of the ladies of Blaenau Ffestiniog happened to include a silk item it would probably be cleaned in a liquid made from boiling grated potatoes in water.

Witness after witness confirmed the town's domestic habits. Dr. Richard Jones testified: "I think the quarrymen are very well clothed when they work. They wear a good deal of flannel – flannel vests and drawers, and stockings – and of course the non-conducting and hygroscopic properties of flannel tell very much in their favour."

Another Richard Jones, a rockman at Middle Oakeley, confirmed the men usually wore flannel next to the skin, saying they wore *trwsus Ffestin,* or sometimes corduroy trousers. *Trwsus Ffestin* (i.e. Ffestin trousers) was a Ffestiniog corruption of *trwsus ffystyn,* meaning fustian trousers, which were virtually the uniform of the Ffestiniog miner. Fustian was a thick off-white cotton material, of a particular weave developed in ancient times at Al Fustat, the original Cairo (now only ancient ruins south of the present city), and introduced to Britain by the Crusaders.

While Dr. Robert Roberts was giving evidence he was asked, by the chairman, if there was want of knowledge of hygiene among the people of Blaenau Ffestiniog. "For instance, if a man wears his shirt for a fortnight or a month he is not likely to be as healthy as if he changed it at shorter intervals?"

"No, certainly not," he replied, adding that they were very conservative in their habits, although he had frequently spoken to them about it.

"Do the quarrymen's wives and their daughters spend too much money on dress – spend money on dress that ought to go into the husband's stomach?" asked the chairman.

"I believe the great curse of the day is that so many cheap articles of apparel are sold and that the quarryman's money goes a good deal in that way. Such articles have to be bought often instead of buying one good article that would last," replied Dr. Roberts.

"I am not throwing stones at the quarrymen specially, but it is a fault one meets with in all classes," said the chairman, adding: "Is it apparent in the quarryman's family?"

Dr. Roberts replied: "I cannot say they are particularly extravagant, but I am sure I could myself prescribe more useful clothing if I were asked. I would require all to wear flannel and woollen and ordinary cotton goods. I would not go in so much for fineries."

John Ellis Roberts, of Tanygrisiau, was questioned at length about clothing. "The general rule is to wear linen jackets and those are left in the quarry at night, and dried by the morning. If the weather is cold we wear old cloth jackets until they are worn out."

"Do you wear flannel?" – "Yes, I always do, and I believe the majority of the men do. The men who are somewhat old always wear flannel, but the younger men may neglect it."

"Do the men change their shirts at night?" – "I do not think that is done."

"Do they wear the same shirt all through the week, day and night?" – "That is how I do, and I believe the great majority of men do likewise."

"Do you think it would be better if they were to change their shirts after returning from their work?" – "I believe that would be an improvement. It would certainly be refreshing. We have our shirts washed as a rule every week or every fortnight."

Dr. R.D.Evans said: "More care in the matter of changing under-clothing and damp clothes would be desirable, and more frequent ablutions would also be of assistance in maintaining the health of the men. A better knowledge of general hygiene, also, would be of considerable advantage. At present we have no facilities for the men in any way to have baths. I consider we ought to be provided with public baths, if we are to progress with the advance of sanitation and hygiene in these days. There is no doubt that a public bath in such a populous district as this would be a great acquisition, and ought to be considered by the Local Board."

Asked how the town's high incidence of rheumatism might be avoided, Dr. Evans said: "By wearing flannel, and that should be kept dry, as far as possible, and changed oftener than is the practice among the men generally. I fear that the general custom is for the men to work in two or three coverings of flannel, and to change only about once a fortnight. The result is that the flannels get cold and clammy, and the effect cannot but be injurious."

Greaves asked: "And he generally goes to bed in the same shirt he works in?" – "Yes."

"And he keeps it on for a fortnight?" – "Yes. I believe it is a matter of habit, and for want of being properly taught. They think it is unsafe to change their flannels. They seem to be afraid of air or water getting into contact with their skin, apart from the hands and face."

Domestic science teacher Eleanor Russell said: "Most of the children do not know what it is to have a proper bath, and of course their health cannot be good without that." She said baths should be provided either in the houses or in a building where they could be used for a small payment.

However Ann Wilson, a teacher at the Girls' Higher Grade School, said all the children she knew used a bathtub on Saturday nights.

Cookery lecturer Winifred Ellis said girls should be taught more about household work. Asked if the wives were extravagant, she said: "There is a lack of thrift, in that

A table of grotesque weight, designed and fashioned by Magnus of London, out of Llechwedd slate, for the Great Exhibition of 1851.Below are prize-winning medals awarded to J.W.Greaves for slate displays at the 1851 exhibition and the Paris Exhibition of 1867. The family also won medals at the London Exhibition of 1862, and the Buenos Aires World Fair of 1910.

they do not make the best use of their materials. Some are extravagant in the cases I have already pointed out. I do not by any means imply they are lazy as a class of people, but some are, and are given to gossip.

"It is my opinion that they do not spend enough time at home. I have heard from very good authority that some of the women go to each other's houses when the men are at work, and when it is time to return they go to their houses and buy tinned meat for their husbands, instead of preparing a nice little dish for them."

Like people who today live near and work at places like the Stanlow oil refinery, who cannot smell the all pervasive petroleum fumes which can sicken strangers passing miles away, the 11,073 people living in the 2,380 houses at Blaenau Ffestiniog in 1893 were presumably inured to the stench of their unwashed bodies – but not to the ever-present smell of their rotting excreta.

It was with obvious pride that Henry Maybury, surveyor to Ffestiniog Local Board (forerunner of the Urban District Council) told the inquiry about the 1881 introduction of what the community called the Midnight Express. This was the community's "night soil service," involving a lorry that went around during the summer nights to collect the human excreta from the houses. "The pails are removed every fortnight, and fresh ones, properly disinfected, are substituted," he said.

Thus for two weeks it accumulated in backyards, and there were places where none was ever removed. Sanitation was a major problem frequently mentioned at the 1893-94 inquiry. John Hughes, one of the strike leaders sacked from Llechwedd, said there were no privies deep underground, and their absence caused a lot of impurity in the atmosphere.

"Are there any privies at all underground?" asked Greaves. "Yes," he replied, "in the Old Floor."

"And are there any on the surface?" – "Yes, there are. The lowest one underground is on the Old Floor." [That was one floor below the surface area now used by tourists, but there were another nine floors below that. The surface privies mentioned by the witness now form an exhibit in the tourist section – the seats are over a stream, which runs down into the town].

J.Lloyd Jones, of Foty and Bowydd quarry, said disused chambers had been set aside for defecation and micturition, and men were fined for using anywhere else.

John Owen, of Middle Oakeley, said there was no privy provision in the mills, and he knew of no such provision anywhere on the surface at his quarry. "Our custom is to go rather far away from where we work," he said, but the habit of nearer urination was a nuisance.

Dr. Robert Roberts said: "I understand the bulk of the excreta of the men is deposited at the quarries and not at home, and this of necessity should call forth proper accommodation where it does not exist, and some means of compelling the men to use the same where it does exist. From the want of this the air in the underground chambers may become vitiated, and the water draining from the excreta may pass through rubbish heaps and crevices and appear drinkable to the men, whereas it may really be polluted with sewage."

Dr. R.D.Evans said: "There are scarcely any privies provided in any of the quarries, with the single exception of Llechwedd. In one quarry, in which 1,600 men are employed [i.e. Oakeley] there are nine privies, and the majority of these are in a filthy condition and scarcely ever used. The excreta from such a large number of men must be enormous and a nuisance, apart from the possibility of the excreta contaminating the water that the men may drink underground, and thus cause typhoid fever, of which I have had experience. In addition to this there is the danger of exposing a portion of their body [by dropping their trousers] when heated, in all weathers, which in my opinion is often the cause of the cases of congestion of the kidneys, and the large number of cases of lumbago and sciatica met with here in the winter and during the prevailing cold easterly winds of the spring."

Oakeley quarry manager Robert Roberts was quick to defend his company. "We have nine privies, which I think, on looking at the question all round, is not enough. Below ground exhausted chambers, away from the working places, are used by the workmen. Above ground the tips, which cover an immense area, are used, and with the heavy rainfall we have, amounting to about 100 inches a year, the excreta are quickly washed off, so that practically we suffer from no nuisance."

William Williams, of Minffordd, working at the New Welsh Slate Quarry, said it would be beneficial to health if proper privies were provided underground. "When there is an old disused chamber the men go there, and that makes the air impure. Pails should be provided, or a hole with water in it, so that they may be cleaned periodically."

In the midst of all this Robert Roberts, a rockman of the New Welsh Slate Quarry, complained: "The men get candles which have such a bad smell that they cannot eat at dinner time." These they bought from the company.

Owen Francis Williams, a rockman at Abergynolwyn, said there were no privies at the quarry, where 200 men were employed. There were some privies in the village barracks used by some of the men.

Merioneth County Councillor Morris Thomas submitted a note suggesting the government's Inspector of Mines should be invested with authority to control sanitary matters.

To put the Blaenau Ffestiniog problem into perspective, the 1895 report was submitted to Parliament in the year that another survey, for a typically large English city, Leicester, revealed that while 13,000 houses and commercial premises had water closets, the remaining population relied upon 6,700 buckets, 11,000 ash-pits and 17,000 ash-bins. Even London did not have a sewerage system until 1875. As recently as 1939 more than half the working-class population of Manchester relied upon dry-ash closets.

A Victorian street scene at Blaenau Ffestiniog, nestling beneath the rocks that created its economy.

Much was said about the miner's propensity for tea-drinking, but all aspects of his diet were worrying those concerned with his welfare. "Great ignorance prevails as to the value of soup as food," said Winifred Ellis, who had lectured on cookery and home economy at classes funded by the County Council at four quarrying centres: Ffestiniog, Harlech, Corris and Abergynolwyn.

"I made a highly nourishing food from 8 lbs of meat, and offered to sell it at a very moderate price, but did not sell a pint of it, and was obliged to give it away. However the meat from which it was made was in great demand although I gave instruction and carefully showed I was making the soup in such a way as to extract the flavour and nourishment from the meat into water.

"I might mention that I was told, but cannot say whether it is a fact, that some of the people roast a joint for Sunday, and serve it hot, but that they rarely think of eating it cold or making it up for another day, although perhaps half of it is left.

"It is a common thing among quarry people to have their [monthly] wages paid on Saturday morning and go to the shops on Saturday night and spend the best part of them, live in comparative luxury for a week, and for the next three weeks live very scantily until they receive another month's wages. I think it would be a wise plan to pay wages weekly."

Miss Ellis said reform of cooking for the sick room was much needed [and there was nearly always somebody sick in every household]. "Many people have but little idea how to make beef tea, mutton broth, and other essentials to the diet of a sick person. It can hardly be expected for an invalid to make progress without nourishing food and careful nursing. Undoubtedly, scores of people suffer incalculably from neglect in these two points," she said. In some houses a sick person lay in a room for weeks or months, breathing the same air because the windows either could not be opened or never were opened.

The town's women did not make the best of the food available, and many men were poorly fed. Asked if the women could implement Miss Ellis's recommendations at the same cost, she said: "It might cost a little more, but they might turn it to better purpose. In cooking a joint they might make it go very much further than they do. They might cook it for Sunday and if they are not a large family a joint might provide for three dinners, or even four."

Committee member John Evans interjected that Blaenau Ffestiniog had a lot of old bachelors because the town's women were so extravagant. "Are the wives of the quarrymen rather backward in their mode of living?" he asked. "In some respects," replied Miss Ellis.

"Are they more backward than the wives of other classes of workmen?" – "What classes?"

"The wives of farmers and joiners, and so on." – "No, they are not, when you come to compare them," said Miss Ellis. She said they did not consume enough fresh meat or milk, and were not very partial to vegetables.

Ann Wilson said she taught cooking at the Higher Grade School and also at evening classes. All were taught various ways of cooking butcher's meat, cold meat, bread-making, soups, vegetables, puddings, cakes, pastry, fish and for the sick room. She suggested children should have a daily dinner cooked for them in school, for a cost of 2d (i.e. less than 1p).

"Some of the workmen only get a proper dinner on Sundays. The other six days they take tea in the quarry, and when they reach home in the evening they have tea again, with bread and jam, or some tinned meat. Each day a proper dinner ought to be supplied to every workman, and some means should be taken to bring this about. A small kitchen might be erected at every quarry, and a woman paid to prepare soup for the men's dinner, at a cost of about 1d each daily. The men should take bread with them to eat with the soup," said Miss Wilson.

She handed the Departmental Committee a recipe for eight gallons of wholesome soup for 80 men, which she said would cost a total of six shillings (30p). It comprised fresh bones, peas, lentil flour, carrots, turnips, onions, salt, pepper and mint.

Miss Wilson said the local women needed to be taught how to make soup, and should be discouraged from using tinned meat. They did not know what to do with the cold meat and bones left over from the weekly joint.

Eleanor Russell, a cookery teacher at the junior schools, said: "If stew were taken at the midday meal, or Irish stew, it would be far better than tea and bread and butter, and they could have a proper dinner in the evening after returning home."

Generally speaking the water used for cooking meat by the women of Blaenau Ffestiniog was thrown away, instead of being used for soup. She thought it would a good idea to set up soup kitchens in the schools and the quarries.

Starting when they were about nine or ten years of age, all girls were taught a wide range of cooking skills but there was no classroom kitchen in which they could develop what they were being taught. The evening classes for older girls were poorly attended.

Dr. Robert Roberts said: "There is a great want of domestic thrift and economy in the district. There has not been sufficient support given in the neighbourhood to such classes. Mrs. [Marianne] Greaves did a good deal in that direction some years ago, and she would no doubt have done a great deal more, if what she proposed had been responded to, as I consider it ought to have been. Even now, from what I hear, I doubt very much whether the cookery classes under the county council are having the support they deserve."

He was referring to the initiative of the wife of J.E.Greaves who, in 1883, invited Fanny Louisa Calder to visit her, at Bron Eifion. Miss Calder was the founder, in 1875, of the Liverpool Training School of Cookery, in Colquitt Street (which became the F.L.Calder College of Domestic Science when taken over by the city's Education Committee in 1921). J.E.Greaves recorded the visit in his diary for 4 October 1883: "Meeting of ladies re cooking classes. Miss Calder present." Just over a year later, on 17 November 1894, he recorded: "Polly [his pet-name for his wife] opened the cookery classes." Thereafter Mrs. Greaves extended her classes into 48 different centres.

Dr. Richard Jones said the men needed more nitrogenous food. "At present they get none, except perhaps what is contained in the milk they use with their tea. They ought to get a good deal more than that, and they should also eat more fat." Oatmeal porridge would be beneficial, and milk or buttermilk would be better than tea, which caused various stomach diseases.

Dr. R.D.Evans did not believe the people lived quite so sparingly as was generally supposed. "I do not think there is a town in North Wales of similar population in which the inhabitants eat so much meat and drink so much milk. On an average from 1,800 to 1,900 gallons of milk are sold here every week, and not far from 15 tons of meat."

THE WELSH LANGUAGE 7

Blaenau Ffestiniog remains a predominantly Welsh speaking town after a century of emigration rather than immigration, which has reduced the population by half over that period. At the 1893-94 inquiry several witnesses suggested the need for a Welsh-speaking deputy to the Government's Inspector of Mines.

The point was first raised at the Departmental Committee's third session, on 20 December, by slatemaker Owen Rowland Jones, from the Oakeley Middle Quarry. "We ask the Government to appoint a sub-inspector possessed of practical knowledge, and a knowledge of the Welsh language," he said.

"Is it not a fact that a great many of the men understand some English?" asked Dr. Foster.

"No, not very many," replied O.R.Jones.

Evans asked: "Do the men, as a rule, find a difficulty in speaking to Dr. Foster? [in his capacity as Inspector of Mines]. – "Yes, that is certainly a disadvantage. The [owner's] Agent is the interpreter as a rule between Dr. Foster and the men, when he visits the quarry."

"And you think that is not quite fair?" – "I believe it would be better if he could ask his questions in the language of the men."

Greaves asked: "What part of Dr. Foster's work at present is not satisfactorily done?" – "Want of knowledge of the Welsh language is one."

"Is there any other suggestion?" – "No, not as to Dr. Foster. Personally, I have no objection to him at all."

Richard Griffiths, who had worked as both a rockman and a slatemaker, was asked by Jenkins: "Do you think it would be of advantage to have a sub-inspector under Government, and that he should be a Welshman and a practical man?" – "Yes, I believe it would."

"You think he should be a man who can speak to the workmen in their own language?" – "Yes, I do, and that for many reasons. In the first place we consider that the quarries require more supervision as they get deeper, and we want an experienced man to see they are properly looked after. We believe the present inspector cannot, by reason of his district being so extensive, see that what ought to be done is done."

Greaves asked Dr. E.D.Evans whether the distribution of pamphlets on first aid would be useful in the quarries. "Yes, if they are prepared in Welsh," he replied.

Union leader D.G.Williams was asked if the men approved of the way Dr. Foster did his inspections. "They approve, as far as it goes. They consider it would be an advantage if he were more practical and understood Welsh," he replied.

The body of the Victorian miner might have been polluted, for the want of what we now know to be fundamental water and sanitation facilities, but his intellect was sound and productive. The town's first printing press was set up in 1861, and in 1877 William L.Roberts began publishing the town's first weekly newspaper, the Welsh-language *Y Rhedegydd* (which continued until 1951). In 1903 Lewis Davies launched his rival weekly, *Y Gloch*. There was an early flourishing debating society, called Senedd Bethania (i.e. the Bethania Parliament), centred upon Bethania Independent chapel, which dated from 1817. Future Prime Minister David Lloyd George became a member in 1885.

An eisteddfod (a Welsh-language literature and music competitive folk festival, dating from the Dark Ages) had been a feature of the area ever since August 1854, when an eisteddfod was held at Llan Ffestiniog. The president was the famous poet Gwalchmai (the Reverend Richard Parry, of Llandudno). That became the basis for an annual Whit Monday eisteddfod, which later gave way to an annual Blaenau Ffestiniog Christmas eisteddfod. In 1868 J.W.Greaves founded the annual Llechwedd eisteddfod for his own workforce, and maintained his interest in it long after he had retired, travelling over from Warwickshire to his second home at Aberglaslyn every August until 1879, to enable him to preside over the event. His sons John and Richard carried on the tradition. Oakeley quarry followed suit in 1879, and the Cwmorthin, Welsh Slate and Holland mines and quarries subsequently introduced their own eisteddfodau. All this activity led to the establishment of *Yr Eisteddfod Dalaethol Gwynedd* (Gwynedd regional eisteddfod) at Blaenau Ffestiniog in 1891. In 1896 the town sent a deputation to Llandudno, then hosting the National Eisteddod, inviting the 1898 festival to Blaenau Ffestiniog. They won the day, against opposition from the very big Liverpool Welsh community.

The 1898 National Eisteddfod was held over five days in July, under the patronage of Queen Victoria, reinforced by such names as the Duke of Westminster, Lord Newborough, John Ernest Greaves, Osmond Williams, Clement Le Neve Foster, Charles Holland, Sir Herbert Oakeley, W.F.Inge, A.J.Balfour and H.H.Asquith. Greetings were received from Queen Elisabeth of Romania who, while staying at Llandudno in 1890, had been admitted to the Gorsedd, during the Bangor National Eisteddfod, under her pen name of Carmen Sylva. It was an event that brought the first royal visitors to Blaenau Ffestiniog: HRH the Duke of Cambridge, grandson of King George III, accompanied by the Prince of Saxe Weimar. He came as chief personal ADC to Queen Victoria, acting on her behalf.

Then, as now, the chair was the principal award, and it went to a local poet, R.O.Hughes (Elfyn), of Llan Ffestiniog, while the crown went to the Reverend R.Gwylfa Roberts, of Portdinorwic – the harbour for the Llanberis slate quarries. Diverting from tradition, the eisteddfod included a competition for slate splitting, which proved very popular – more especially when all the prizes went to local men.

Our erstwhile Dr. Robert Roberts is credited with having revived the peculiarly Welsh *penillion* tradition, in which the vocalist sings a poem to a different tune to that played by a harpist accompanist, but both finishing at the same time. It was a tradition dating back to the dark ages when minstrels roamed the country earning a living by singing the news of faraway events. The eisteddfod began as a contest for sorting out the true wandering minstrels from the outlawed vagabonds.

Himself an approved poet and harpist of the Glan Geirionnydd counter-eisteddfod, held in the Conwy Valley, where he had been given the bardic name of Isallt (derived from his ancestral home), Dr. Roberts was elected chairman of the Musical Committee for the 1898 National Eisteddfod. He was living at Plas Weunydd, Llechwedd, in 1913 when he set off for that year's National Eisteddfod at Y Fenni (Abergavenny) to deliver a notable lecture.

"The interest displayed of late by the supporters of the time-honoured and unique method of singing *penillion* with the harp, and their desire to restore it to its old repute, has now increased to such an extent that there is every reason to believe the whole Principality, from Mona to Monmouth, is ripe and ready to assist in raising the art form from the low level of neglect it has been allowed to drift into, and to re-crown it on the throne of popularity," said Dr. Roberts.

"Most of the leading *penillion* singers entertain personal views and opinions on the subject. These are so varied and diversified, and the rules are in such a chaotic state, that a master hand is required to restore them into some order more consistent with the present-day requirements of the literature of music, of which the art of *penillion* singing may be looked upon as a branch. It is generally held that there is something indefinable in the music of the strings *(gwawd y tannau),* which falls into sweeter and deeper harmony with the inner nature of the Cymro (Welshman) than any other kind of music," added the doctor, then aged 74.

He told the influential audience that more harps were needed for Welsh children before a general efficiency in playing could be attained. There was also a shortage of teachers and Dr. Roberts suggested a solution that took several more decades to materialise – peripatetic teachers, visiting a chain of centres where the pupils could be regularly assembled. He advocated arrangements for every pupil to be supplied with a Celtic harp *(y delyn fach),* on account of their cheapness – between £2 and £3. He said the fingering was identical with that of the pedal harp, and he hoped that as pupils excelled, their parents or friends would procure the ordinary-sized harp, "and thus Wales will, ere long, be stocked with harps as of yore, *telyn ym mhob teulu* (a harp in every family)," he added.

Dr. Roberts offered a long list of suggestions for improving the conventions of *penillion* playing. "The present chaotic state of the rules of competition, and the dissatisfaction prevalent amongst competitors, demand that the adjudicators should follow some recognised authority. The art must also be raised to a position of greater dignity in the opinion of the audience," he said, adding: "It is high time to wipe out the wrong impression that *penillion* singing is only an item put into the programme in order to create merriment."

The Welsh language issue in Blaenau Ffestiniog went far beyond the sessions of the 1893-94 Departmental Committee. Alexander Murray Dunlop, general manager at the Oakeley Quarry, stood as Tory candidate for Merioneth at the 1880 general election, polling 1,074 votes to Samuel Holland's 1,860. Two years later Dunlop switched to the Liberals, and it was under that label he was elected a founding member of Ffestiniog Urban District Council, which first met in January 1895.

Other members of that first town council included several people who had been prominent during the sessions of the Departmental Committee, including John Jenkins, Dr. R.D.Evans, Charles Warren Roberts, Robert Roberts and William Owen (not the Llechwedd striker, but one of the managers at Oakeley, who, in 1887, married a niece of Dr. Robert Roberts. He was appointed manager at Llechwedd in 1897, and moved into Plas Weunydd). The first item on the agenda was the election of a chairman. George Ellis and Cadwaladr Roberts proposed Dunlop, who had been the last chairman of the previous Local Board, but the nomination was opposed by John Morgan and Evan T.Pritchard, who said the job should go to a Welsh speaker. The Welsh language lobby won the day and Robert Roberts, Dunlop's deputy at Oakeley, was elected [He became the father of three famous doctors in the North Wales slate quarrying areas: Robert Herbert Mills Roberts, Richard Arthur Mills Roberts, and Edward Mills Roberts].

The council later turned its attention to the London & North Western Railway Company, which had dismissed several of its Blaenau Ffestiniog staff only because they were monoglot Welsh speakers. The council called on local traders to boycott the LNWR until such time as this injustice was put right.

The Government responded to the case for a Welsh-speaking sub-inspector by appointing schoolmaster Griffith John Williams an Assistant Inspector of Mines in 1895. Presumably it was thought his service as interpreter for the 4,786 questions asked by the Departmental Committee had taught him enough about the practicalities of the slate industry. He died in 1933, aged 79. Owen Rowland Jones, who had first raised the language issue at the 1893-94 public inquiry, was also later appointed an Assistant Inspector of Mines.

The Greaves family pioneered the electrification of the slate industry – an historical fact rival 21st century enterprises are reluctant to acknowledge. On the sixth day of the Departmental Committee's inquiry (17 January 1894), when Llechwedd mine manager Charles Warren Roberts was talking about incline operation, J.E.Greaves asked: "Are any means adopted for lighting the incline?"

"Yes, in many places there are electric lights, especially where the wagons are hooked on to the ropes," said Roberts, adding that this greatly increased safety. He said Llechwedd mines used electricity for many purposes: lighting, pumping, rock-drilling and fan-driving, both underground and in the surface sheds, the carpenters' shop and the office, using water that formerly ran to waste to turn a turbine.

We know from Ffestinfab's book *Hanes Plwyf Ffestiniog* (History of the Parish of Ffestiniog), published in 1879, that Llechwedd had by then experimented in the use of electricity for haulage, as a replacement for shafted and geared power derived from enormous water wheels, but the electrical equipment available in the 1870s was ineffective. It was not until 1881 that Britain's first electricity generating station was built (at Godalming).

The inspiration for the innovation and practical introduction of electricity to North Wales came from Richard Methuen Greaves, brother of J.E.Greaves, who had trained as a marine engineer with Byer & Peacock and the De Winton Union Works, Caernarfon, before joining the family business. He made his home at Wern, near Porthmadog, first as tenant but buying the entire estate in 1886. Three years later he equipped the house with electric lighting, generated from his own water supply. He installed a James Thomson variable speed vortex turbine made by Gilbert Gilkes, generating 80 volts and a maximum load of 45 amps (3.6kW), at 1,220 rpm. The equipment remains in situ, and was used into the 1980s to drive a forced-draught fan for an oil-fired kiln. His brother's diaries tell us that on 14 April 1890 they both met at Llechwedd with a Mr. Rooper, representing Gilbert Gilkes, to order similar equipment for the quarry. Family diaries tell us that equipment was up and running for its first test run on 9 October. It was the first industrial electrical installation in North Wales. The pioneering dynamo was located beside the Floor 2 mill. J.E.Greaves installed a hydroelectric set at his new home at Glangwna, Caeathro, on 9 December 1892.

By 1896 the Llechwedd quarry installation had been upgraded to a single Thomson vortex turbine driving two dynamos of 120 amps and 60 amps. Owning two lakes, Llyn Bowydd and Llyn Newydd, high above Llechwedd, the brothers fully harnessed this resource on 11 April 1904, when the present Llechwedd-owned Pantyrafon powerhouse was brought into use, at the foot of the main incline, which passes beneath the A470 road. It was equipped with two Pelton wheel 250kW dynamos, made by Gilbert Gilkes, each of 550 volts. They remain in daily use, via rotary converters to create alternating current, sometimes generating a surplus that is fed into the national grid. The rotary converter had been available since 1892.

Left: An early electrical installation (left) at Llechwedd. The dynamo is belt-driven from an enormous water wheel. The switchboard was made by A.Lester Taylor, of Liverpool. One meter is registering a delivery of 124 Volts, but no load is shown on the adjacent ammeter.

At the 1893-94 public inquiry the chairman asked Warren Roberts: "Do you think the use of the electric light in the quarries would tend to safety, for instance, that with plenty of light at [railway] pass byes and sidings, and inclines, accidents would be avoided?"

"I have found it very advantageous at Llechwedd," said Roberts, "because it enables the men to be quite sure they have placed the chain around the block properly. They get a very bright light to work in. Unless the block is securely fastened it is very liable to cause an accident when going up the incline."

Asked whether electric light could be used in all quarry work, Roberts said he did not advocate it for working the rock face, where the man needed to put his candle at any point so as to get the light exactly where he wanted.

E.Parry Jones asked for some information about the cost of installing and running electricity in a slate mine. Roberts replied: "Given an available water supply you can fix up a plant, say of 25 h.p. for the sake of illustration, at half the cost of obtaining the same horse-power by steam." A water supply was the prerequisite, but the electricity supply could be distributed over a large area with a very small loss, compared with compressed air [for powering the Moses Kellow pneumatic rock drills], transmitting with an efficiency of 70% compared with 25%.

His experience of three years of electrical operation at Llechwedd was that maintenance costs were very small. "The only wear has been on the commutator [*sic*] of the dynamo, and that not a 32nd part of an inch." [Today's electrical engineers reserve the term "commutator" for the device used to produce alternating current from an alternator, in place of the "slip-rings" to tap direct current from what then becomes a dynamo, although otherwise the same machine].

In response to questions, Roberts said hundreds of horsepower were running to waste every day from the abundance of water available in the area.

Evans asked: "Is electricity quite as manageable a power as any other?" – "Certainly it is. It can be managed easily; a man does not want a special training to look after the electric machinery in the quarry. Of course you will need an expert now and then to renew something, or to do any particular repairs, but for the ordinary working of it you want no more skilled labour than you do for the engine-driver, and not so much."

"And you can use it for sawing purposes?" – "Yes."

"And for winding purposes?" – "Yes, for any purpose you like."

As early as September 1891 Roberts had tried to persuade the Local Board to introduce hydroelectricity to Blaenau Ffestiniog and Tanygrisiau. The matter was investigated, and among the proposals received was one from Gilbert Gilkes, but the matter was not pursued. When the Local Board was replaced by the Urban District Council in 1895 Roberts again raised the issue, but it was three years later before the local authority asked for a feasibility report. At that time 125 street lamps and 245 houses were lit by gas, and the council invited tenders for electric replacement. There were 34 responses but the council deferred implementation until May 1902, when the electric street lamps were formally switched on by Lord Newborough, freeholder of much of Blaenau Ffestiniog – seemingly Britain's first hydroelectric public supply.

In 1894 Charles Warren Roberts, of Plas Weunydd, Blaenau Ffestiniog, and Edward Henry Beckett, of Glaneulan, Tanygrisiau, were registered by the United States Patent Office as inventors of the world's first electrically operated weighing scales (their signatures being witnessed by the Reverend David Richards, Vicar of Blaenau Ffestiniog, and Richard Jones, of Brynmarian).

In 1899 the Votty & Bowydd Slate Quarries Co. Ltd. created the subsidiary Yale Electric

Engineer Richard M.Greaves behind one of his pioneering dynamos (above), here used for battery charging at Llechwedd Slate Mines. It was rotated by a short shaft linked, by belt drive (right), to an overhead shaft turned by a water wheel. The drive wheel is all that now remains of this early generation system. Modern tourists can see it in the demonstration mill at Llechwedd Slate Caverns.

Power Company (taking its name from the managing director, Osborne Yale, who died at Betws-y-coed in 1923). Using the water of lakes Conglog, Cwmcorsiog, Stwlan and Cwmorthin, to feed two 9kW dynamos at a power station built at Dolwen, Tanygrisiau (some two miles south of Blaenau Ffestiniog), the company began generating at 500 volts DC, in 1900. In 1902 the Yale company upgraded their system with two 90kW dynamos, operating at 500v. In that year the Great Western Railway station at Blaenau Ffestiniog became the first station in the company's Chester division to be lit by electricity, supplied by the Yale company.

Originally intended for supplying power only to quarries, Yale Electric began supplying domestic customers with 230v, adding to their generating capacity in 1905, 1926 and modifying the plant in 1937 to replace the DC generators with 589kW alternators, at 400v, with an annual output of 3million kWh. It continued to supply all the houses and shops until the 1948 nationalisation of the generating industry, when the town became part of the national grid. Dolwen Power Station was switched off in March 1964.

In 1901 Moses Kellow, who had sat through the 1893-94 inquiry, introduced electricity to Croesor Quarry, using water harnessed from Cwm y Foel, initially to turn a dynamo, but soon replacing it with a 440 volt three-phase alternator, imported from Prague, to give a 240 volt distribution. Cwm y Foel dam was deliberately breached in 1986 to release the water back into the Croesor river. Alternating current had been available for several years but at that time there was a transatlantic debate as to the most efficient voltage for transmission by copper cables. When Thomas Edison proposed supplying electricity to American buildings, some of which were then using gas lighting, the powerful gas industry took him to court, claiming electricity was too dangerous to be supplied to houses. The court ruled the safe level to be 100 volts, and allowed a variation of plus or minus 10%. Edison immediately absorbed the 10% margin and introduced his 110v supply that remains the standard across the United States, although there are now a few 240v installations. For his slate quarry in Merioneth, Moses Kellow adopted the more efficient 240v that was by then becoming the European standard.

North Wales Power & Traction Company was set up in 1904 to buy the water rights of Snowdonia from Gwalia Ltd, formed in the late 1890s. North Wales Power installed turbines to supply electricity to three major slate quarrying areas, using four 1,000kW generators and 11kV transmission from its Cwm Dyli station, driven by the water of Llyn Llydaw. *The Slate Trade Gazette* noted, in October 1905, that 11kV was many times greater than the power the Board of Trade had hitherto regarded as safe for overhead wires. It was on 13 August 1906 that the company began its commercial supply of electricity.

"One of the lines goes to Ffestiniog, across the Lledr Valley, there to supply the Oakeley Quarry; another to Llanberis, via Pen-y-pass to supply the Great Dinorwic quarries; and a third to Nantlle Vale, via Drws-y-coed Pass, to supply Pen-yr-orsedd Quarry, and probably later on several more quarries along that Valley," noted the *Liverpool Daily Post,* adding that electricity would reduce the quarry coal bills by 30-40%. Conservation of the natural beauty of Snowdonia was already a sensitive issue, and the *Daily Post* noted that Cwm Dyli power station had been designed to look like a church. It was quickly dubbed "the Chapel in the Valley," and remains an attractive building in its majestic mountain setting.

As well as supplying the slate industry, North Wales Power was allowed, by its founding Act of Parliament, to supply power to anywhere in Caernarvonshire, Anglesey, Merioneth and part of Denbighshire (eventually extending to Crewe). In many towns that resulted in a situation where houses received a 15 amp or 30 amp supply from North Wales Power, and a separate 5-amp lighting circuit from the local council, monitored by different meter readers.

Still in situ – and still in working order – Richard M.Greaves's Victorian low voltage hydroelectric generation installation at Wern, his home near Porthmadog – now a nursing home.

In 1912 the North Wales Power Company extended their distribution to Penrhyn Quarry, in the Ogwen Valley, where a 750kVA 20,000/500 volt transformer delivered a 500v three-phase internal distribution.

It was the existence of the Cwm Dyli power station, built for the Welsh slate industry, that influenced Guglielmo Marconi when choosing a site for his first transatlantic radio transmitting station, which he began building at Waenfawr, near Caernarfon, in 1912. North Wales Power provided his Marconi Wireless Telegraph Co. Ltd. with a 30kV three-phase supply, for 440v use in the long-wave transmitter which began broadcasting in 1914.

In 1928 North Wales Power dammed up the poet Hedd Wyn's famous Cwm Prysor, to create Trawsfynydd lake, with which to drive Maentwrog hydroelectric station, just $2^1/_2$ miles from the home of electricity pioneer R.M.Greaves (who lived until 1942). The station was designed to look like a school, and was originally equipped with three generators capable of producing 18 megawatts – when the maximum demand for the whole of North Wales was 13.3 megawatts. The existence of Maentwrog dam was one of the factors that resulted in the shores of Trawsfynydd lake being chosen in 1953 as a site for a nuclear power station. The dam was strengthened and increased in height and the nuclear power station began generating 230 megawatts in 1965. It reached the end of its active life during the 1990s, and was closed down in phases.

In 1955 Stwlan lake, which had formed the basis for the 1899 Yale scheme, attracted engineers looking for a site for Britain's first hydroelectric pumped storage scheme. Work began in 1957 and the Queen opened the Tanygrisiau power station in 1963. Known as the Ffestiniog Pumped Storage Scheme, it has four 90 megawatt generators, feeding 275kV into the national grid. Acting like a battery, it can insert 360MW into the system in 55 seconds. In off-peak hours the generators become motors, to pump the water back into the Stwlan reservoir for re-use.

Cwm Dyli power station, here seen nearing completion, is still called the Chapel in the Valley, having been carefully designed to mask its industrial purpose in an environmentally sensitive area of Snowdonia.

Photographs collected in 1903-04 by future station superintendent William Hwfa Williams, during the construction of Cwm Dyli power station, which was built to supply electricity to the Welsh slate industry.

J.W. Greaves's invention for a slate-sawing table, now among the equipment preserved as part of the Victorian Slate Heritage exhibition at Llechwedd Slate Caverns.

THE COMMITTEE'S CONCLUSIONS 9

We learn from the diaries of J.E. Greaves that the Departmental Committee met for the last time on 29 October 1894, to endorse and sign their long report, which went to the Home Secretary on 10 December, and was presented to Parliament in 1895.

Plunging straight into the geology of the region, the committee recorded a good technical description: "The slate of Merionethshire is of Silurian age. It occurs in thick beds, locally called 'veins,' which are interstratified with beds of felspathic ash, generally fine-grained, but sometimes becoming coarse agglomerates. They are known as 'hards,' and are of considerable importance in furnishing a solid roof for the slate workings. The thickness of the beds worked varies from 30 to 126 feet. In the Ffestiniog district the beds are dipping at angles varying from 15° to 35° to the north and northwest, but in other parts of the county the dip of the strata may reach 70° or more. The dip of the 'split' or cleavage is usually greater than that of the stratification; at Ffestiniog there is a difference of about 15° between the two.

"In addition to its fissility along the cleavage planes, the slate has the property of rending, without great difficulty, along planes at right angles to the cleavage planes. This property of rending or 'pillaring' is of the greatest importance to the quarrymen, for it enables him to bring down large masses of slate intact, with a comparatively small expenditure of labour, and it facilities the sub-division into blocks suitable for making the various sizes required for the market. The 'pillaring' likewise regulates the direction given to the upright masses of slate rock, locally called 'walls,' which are left for supporting the overlying strata. Besides the cleavage and the 'pillaring,' the slate-getter has to consider the natural joints, which traverse the bed of slate in various directions."

The committee reported on the method of working the slate, saying: "Fifty years ago the slate in the Ffestiniog district was worked in open quarries, but as their depth increased the amount of worthless 'top,' 'cover,' or 'overburden' became so great, that it seemed preferable to resort to underground mining. After a time the managers adopted the system now in vogue in the largest mines, which may be described as working by alternate chambers and long supporting pillars, or 'walls,' both of which plunge downwards in a direction approaching that of the dip of the strata. The save in the cost of removing useless overlying rocks has been thought to be sufficient to compensate for the great loss of slate left in the form of supporting pillars, and for the numerous other disadvantages of working the mineral underground.

"Looking at the matter by the light of later experience, it seems not improbable that it would have paid in several instances to have gone on longer with the old system of open quarrying," said the committee.

[J.E.Greaves's diaries tell us he took this suggestion on board, and on 4 September 1894 he and his brother R.M.Greaves began discussing the "untopping" of a large area at Llechwedd. However, the family had to wait until 1934 before the 5th Baron Newborough agreed to accept £8,000 for his inherited right to graze nine cows on Llechwedd common, enabling them to remove the topsoil].

"For assistance in climbing up the rugged slope of the working face (bargain) the slate-getter (rockman) has a chain fastened upon the floor above, which he twists round one thigh while at work. When boring a pillaring hole with a 'jumper' [a very long chisel] he generally stands upon a piece of timber or a plank held up by a couple of iron pegs inserted into holes in the working face.

"As only two or three men are employed in a chamber, the work of removing the mass of slate before them lasts for several years. Cases may be cited in which the same men have worked even ten years in one and the same chamber, and on one and the same floor. As the work proceeds they gradually lose touch with the roof, and finally may find it considerably more than 100 feet above their heads, in fact inaccessible without the use of ladders. The only light used in the actual 'getting' is that of a tallow candle held in a lump of clay."

The massive blocks of slate brought down by blasting were usually too large for handling, and had to be subdivided into blocks weighing from 1cwt to 2½ tons. A chain was put around each block, which was dragged away by a crane and loaded on to a railway trolley, which was trammed to the nearest incline. The rubbish was trammed out the same way, in wagons. Each working chamber was separated from the next by a pillar, or "wall," some 40 ft thick.

Only gunpowder was used as an explosive for extracting the slate blocks, because it would sever the rock without smashing it. Some gun-cotton and the various nitro-glycerine explosives were used for some of the preliminary working, such as driving levels, and cutting the "free side" from which to start making a new chamber.

At the incline the loaded trolley was attached to a wire rope and winched to the surface by steam or water power. It went straight to the dressing mill to be made into roofing slates or slabs for cisterns and billiard tables, in direct partnership with the underground slate-getters [each block marked with the team's symbol]. The slabs were cut into suitable lengths by means of circular saws, on a table invented by John Whitehead Greaves in 1850. The blocks were split into roofing slates and trimmed to give straight sides with what the Departmental Committee described as a "Greaves' machine" in general use throughout the area. This was another 1850 invention of J.W.Greaves, but was replaced in 1886 by R.M.Greaves's improved patent, which was soon to be electrified.

Turning their attention to accidents – the fundamental reason for the Home Secretary's decision to appoint the Departmental Committee – they listed 147 separate accidents that had caused the loss of 163 lives, in the previous 19 years. Most were caused by falls of rock, the problem exacerbated by the lack of lighting other than tallow candles. The annual death rate for the period was 3.23 per 1,000 men working underground, and 0.7 per thousand for surface workers, giving an industry average of 1.94 in Merioneth. At Penrhyn quarry, in the Ogwen Valley, the industry average figure was 0.76, and at Dinorwic quarry, Llanberis, 0.71.

"Considering the very large quantities of explosives used daily, and the fact that much is loose gunpowder, the list of fatalities is not so great as one would expect. The quantity of explosives consumed annually at the Oakeley mines amounts to 60 tons of gunpowder, 14 tons of gelignite, and 5 tons of gun-cotton. On the whole, therefore, it cannot be said the men are careless in performing the various operations connected with blasting," said the committee.

"Without wishing in any way to underrate the risks run by the Merionethshire men it is our duty to point out that they are better off than several other classes of workmen," noted the report. British merchant sailors had a death rate of 10.38 per 1,000, British railway workers 2.27, rising to 5.27 for brakesmen and goods-train guards, and 5.41 for shunters; American railway workers 3.27 per 1,000, rising to 9.52 for engine drivers, firemen and guards. The death rate among those engaged in making the Manchester Ship Canal was 2.48; and for the building of London's Tower Bridge 2.9.

"The evidence brought before us by the medical men was of the utmost importance, for their long experience in the district and their total familiarity with the subject in hand entitle their opinions to the greatest respect.

"Dr.Richard Jones, after examining the death registers of the district, came to the following conclusions:

i. The Ffestiniog quarryman enjoys longer life than the average stone and slate quarrier throughout England and Wales;

ii. The death rate from phthisis in the district is lower among the slate-quarriers than among the non-quarriers;

iii. The death rate from phthisis is higher among those engaged in the slate mills than among the underground workers (rockmen and miners);

iv. Pneumonia is the most common disease from which the Ffestiniog quarryman suffers, and the death rate from this disease is higher among the quarries than among the non-quarriers of the district;

v. The Ffestiniog quarryman dies from accidents at a higher rate than stone and slate quarriers as a class;

vi. Excluding accidents, the mean age of death of quarrymen is higher by one year than that of the tradesmen of the district;

vii. Including accidents, the mean age of death is one year lower than that of the tradesmen of the district.

viii. The Ffestiniog quarryman dies from diseases of the digestive system at a higher rate than males of the same age throughout England and Wales, and also than stone and slate quarriers as a class.

The report noted the three principle health problems of the quarrymen were lung diseases, especially pneumonia; rheumatism; and indigestion. "The doctors considered that the cold damp climate of the Ffestiniog district was an important factor in causing rheumatism and diseases of the respiratory organs; they told us that many of the houses were damp, both from standing on undrained land and from the absence of a damp-proof course or watertight layer between the foundations and the superstructure. They were unanimous in condemning the present barracks on account of overcrowding, dirt and the absence of proper sanitary arrangements.

"The diet of the men is far from being all that the doctors would desire; they take too little nitrogenous food, and drink too much stewed tea. Indigestion and nervous debility are both attributed to the tea drinking, though excessive smoking may likewise account to some extent for the latter form of disease.

"With the nature of the clothing the medical men find no fault, as flannel is commonly worn next to the skin; but they point out that the underclothing should be changed more often than it is, and that more frequent ablutions are indispensable. Wearing the same shirt by day and night for two weeks or more at a time is an unhealthy practice that is not uncommon.

"As an instance of what can be done by improved sanitary conditions, Dr. Richard Jones told us that the introduction of a supply of pure water reduced the number of deaths in the Ffestiniog district from typhus and typhoid fevers from 12.69 per annum during the ten years 1865-74, to 1.3 per annum during the ten years 1880-90. Sir George Buchanan and others have proved that the death rate from consumption is immensely reduced in towns by the introduction of a proper system of subsoil drainage. [Sir George Buchanan was Principal Medical Officer for the Government's Local Government Board, 1879-92, and chairman of the Royal Commission on Tuberculosis].

"Respecting the influence of the employment upon the health, all three doctors agreed that the pneumonia, which is so prevalent among the underground men, is largely due to the chills contracted during work or during journeys to or from Ffestiniog.

"According to Dr. Jones, the death rate from phthisis of those who work in the mills is higher than that of those who work underground. He does not appear to be able to say how far this is due to the nature of the employment and how far to their bodily condition on coming to the mine/quarry. The men in the mills are usually brought up in the quarry villages, and begin to work at the age of 14; many of the underground workers, on the other hand, are recruited from farm labourers, who do not leave their healthy employment in the fields until they are 18 to 25 years old. Consequently the underground worker is in many instances better armed for the fight against disease than the surface worker. This fact alone may account for his living longer, in spite of his breathing an atmosphere often polluted with powder smoke. Further, it is suggested that the slate-maker suffers from dust produced by the various operations in the mill. The men believe that they do unquestionably suffer from it, but the doctors, in the absence of any direct proof by post-mortem examinations, are not prepared to admit as a certainty that consumption has been caused by the dust of the Ffestiniog mills. They all, however, consider that there is enough dust both above and below ground to be injurious to the health of the men; to what extent the injury may go, they are unable to say. Dr. Richard Jones, in one of his tables, shows that there is ample air space in the mills, viz. 5,019 cubic feet per man on an average, whereas 250 cubic feet are considered to be sufficient by the Inspectors of Factories under ordinary circumstances, and 400 feet when the men work overtime."

Only two quarries had hospitals, Oakeley and Llechwedd, although men injured at other quarries had occasionally been taken into the Oakeley hospital for a fixed maintenance free. The three doctors were of the opinion that the men who had to travel by train, especially on the narrow gauge Ffestiniog Railway, suffered for want of better arrangements in the carriages and on the stations. "According to Dr. Evans there are 222 men travelling daily by the narrow gauge line in carriages which are much overcrowded, for each man has only about $15^{1}/_{2}$ cubic feet of space; these carriages are ventilated by means of three windows of 19 inches square on each side, which open behind the nape of the neck of the passengers; consequently the men have either to sit in a draught and risk taking cold, or to breathe a very polluted atmosphere."

The report to Parliament noted: "Mr. C. Warren Roberts, of Llechwedd, who has already introduced many important improvements in the mine under his charge, gave valuable and interesting evidence. We learnt from him how power is being transmitted by electricity in his mine for the purpose of working pumps and ventilating fans, and supplying light along the inclines, as well as in the fitting shops, etc., above ground. The safety brake, invented by him and Mr. [Edward Henry] Beckett, seems capable of preventing many of the accidents which happen with cranes. Up to the present time he has not perfected his ingenious appliances for sounding and examining the roof of a high underground chamber without the aid of ladders."

The committee noted the suggestion by Meyrick Roberts, manager of Bryneglwys Mine, Abergynolwyn, that a five-day working week, instead of five and a half, was better for everyone as the mine owner also owned the whole village, enabling the men to rent one of his horticultural allotments to find a healthy and profitable form of recreation on Saturdays.

They were impressed by a set of suggestions placed before them by a committee of workmen:

(1) No work should be commenced in a new chamber until the inspectors of the company certify it is safe to work in;

(2) Periodical inspections should be made of every chamber in the mine, and a record of their condition kept in a book accessible at any time to the Government inspector;

(3) Water should be prevented from trickling down the working face;

(4) Walls should be built up across disused levels to prevent stones being thrown by blasting from one chamber into those adjoining it;

(5) Every crane should be provided with an efficient brake;

(6) A sub-inspector having a practical knowledge of quarrying and speaking Welsh should be appointed;

(7) The inspector should receive accounts of the accidents from the workmen themselves, without being dependent upon the agents as interpreters;

(8) Suitable eating houses should be provided;

(9) A proper supply of drinking water should be available;

(10) Proper privies or closets should be provided;

(11) Some of the persons summoned upon coroners' juries to investigate mine accidents should have a knowledge of mining.

Other suggestions made by witnesses, and which impressed the committee, were listed as:

(a) The mine owner should appoint a person to examine the sides of every chamber each time an inclined slice (thickness, or tew) is worked off, and see that the sides are left safe;

(b) It would be an advantage if a man had had some experience as a slate-maker before working underground;

(c) No man who has not had some experience under the supervision of a practical rockman should have charge of a working place (bargain) or of blasting operations;

(d) Pillars should not be cut away beyond a certain limit, except when a part of the mine is going to be abandoned.

"At Rimogne, we were fortunate enough to be conducted through the underground workings of the St. Quentin pit by M. Lemmens, the managing director.... The bed which is being worked is from 25 to 50 ft. thick, and dips at an angle of 45°. The system of working bears some resemblance to that employed at Fumay, that is to say, the pillars are formed along the strike and the bed is worked off in slices from beneath upwards, beginning at the lowest layers, whilst the workmen stand on the rubbish and have all the overhanging rock within easy reach. M. Lemmens informed us that accidents from explosives had become almost unknown since they had adopted cartridges of compressed gunpowder in place of the ordinary loose gunpowder formerly in use.

"Though not suggesting that this mode of working is applicable to the Ffestiniog district, it yet possesses the following advantages: (1) Smaller amounts of waste material conveyed to the surface; (2) General collapse of the workings rendered impossible, owing to the chambers being filled up with rubbish; (3) Roof of the chamber always within reach, rendering careful examination very easy.

"Now that travelling is so easy and inexpensive, we think that the Merionethshire slate mining companies might, with advantage, occasionally send their managers and under-managers to other districts at home and abroad, for the purpose of learning through the eye the improvements which are constantly being made in the art of mining."

Collecting the underground rubble for winching to the surface for disposal on the vast tips that encircle Blaenau Ffestiniog.

In their report to Parliament the Departmental Committee listed their recommendations under four main headings: prevention of accidents, care of injured persons, promotion of health, and "general matters."

"As long as the present method of working by huge open chambers underground is pursued there will be occasional unexpected falls from the roof or from the sides. Something may be done to render falls less frequent by the careful inspection of each chamber, at intervals of not more than six months. This work might be facilitated by using some powerful illuminant, such as an electric search light, and telescopic ladders specially constructed for the purpose."

They thought slate extraction would be safer by cutting the sides mechanically instead of wrenching by blasting. That would make the sides safer and reduce the use of explosives, with resultant accidents and troublesome powder smoke. It would also reduce the enormous waste of valuable slate.

Prohibiting the use of iron and steel tools when charging and tamping drilled holes with gunpowder could reduce some accidents caused by explosives. "Loose gunpowder has been the cause of many casualties; we lean to the opinion that wherever it is practicable the use of naked powder should be discontinued. The practice of firing holes with two different explosives, such as gunpowder and gun-cotton, at the same time has likewise led to accidents, and we think that explosives should in no case be used save in the manner recommended by the makers," recommended the committee. They said men should be forbidden from riding on slate wagons, and every crane should be fitted with an efficient brake.

On the subject of caring for injured persons the committee recommended that every man and boy be taught the best methods of first-aid, for which the owners should provide training facilities, showing personal interest in the classes and offering prizes. Ambulance corps should be established, similar to those existing in some collieries. Slate quarry owners should be compelled by law to provide stretchers, splints and bandages, as was the law in coal mines. "We would even go further than this and not allow any person to be a manager until he had taken the certificate of the St. John Ambulance Society, or some other society of like nature and equal standing." [Ninety-eight years later, in September 1993, HRH the Duchess of Gloucester, Commandant-in-Chief of the St, John Ambulance Brigade, visited Llechwedd during celebrations for the 75th anniversary of the founding of the Priory for Wales of the Most Venerable Order of the Hospital of St. John of Jerusalem].

"The thorough recovery of injured workmen is often retarded or endangered by treatment at home, where, with the very best intentions on the part of relatives, there may be a lack of proper diet and conveniences, and utter ignorance of nursing. We fully recognise the good which has been done at Ffestiniog by hospitals, and we endorse the desire of medical men that hospital accommodation should be provided for the workmen of all the mines. Even if more hospitals are established, and if the prejudice which exists against them in the minds of some persons is overcome, some cases will always be attended to at home. We may therefore fairly insist upon the advisability of teaching every girl the rudiments of the art of nursing. The knowledge is sure to be of service to every woman at some time of her life, if not for accidents, at all events for disease. The movement might be stimulated by establishing in a centre like Ffestiniog a properly uniformed nursing corps of women, ready to render their services in case of need," continued the committee.

Parliament was told there was no difficulty in making suggestions for the improvement of the health of Blaenau Ffestiniog workmen. "The three great requisites for wholesome dwellings are that they should be dry, light and clean; and in a rainy climate like that of the mining/quarrying districts it is necessary to insist, very strongly, upon excellence of construction, and upon the observance of all precautions calculated to prevent damp walls and damp floors. We consider that the local authorities in the quarry districts should see that their towns and villages are properly drained, and should fearlessly condemn all dwellings which do not present the necessary guarantees of dryness and perfect sanitation."

Recognising that the food of the quarrymen, and its mode of preparation, could be greatly improved, the committee nevertheless acknowledged the impossibility of suddenly altering the habits of the community. "Any change must begin with the young, and it may be accomplished gradually by laying more stress upon the teaching of cookery in schools. We consider the present course too short, but if the maximum grant for the subject can be earned in five weeks, the school authorities have no inducement to prolong the teaching. Cookery cannot take the place it deserves in the school curriculum until the Education Department becomes impressed with the fact that it is an art of paramount importance to the working classes.

"The evidence of the doctors was to the effect that the workmen are suitably clothed, but that they do not change their underclothing or perform their ablutions as frequently as is desirable. The blame for the want of cleanliness with which the quarryman is reproached, is partly attributable to faulty barracks, without appliances for washing or living in decency and comfort. The Inspectors of Mines have directed attention to the shortcomings of the barracks on more than one occasion in their reports. The Ffestiniog Local Authorities have at last awakened to a sense of their responsibilities, and, as far as the barracks are concerned, no doubt matters will be improved, though we do not agree with them in allowing two men to occupy one bed.

"For the people generally it is necessary to go further than the rules and regulations which can be enforced by byelaws. They must be taught from their earliest childhood that the body must be respected, that baths are essential, and frequent changes of underclothing a *sine quâ non* if health is to be preserved. Let the local authorities establish public baths and let the rudiments of hygiene be taught to every boy and girl. It is useless to hope for improved hygienic conditions among the workmen until the schools, local authorities, and persons in position act as guides, and point out the paths which lead to health."

The recommendations expressed concern for the health of men travelling by the Ffestiniog Railway. "The men waiting for their train, sometimes in wet clothes, may often be seen sitting on the ground at one of the stations... as there is not sufficient bench accommodation. A journey of three-quarters of an hour in wet clothes in a draughty railway carriage may easily lay the foundations of disease, and quarrymen who can live within easy walking distance of their homes would, *a priori*, be expected to be less subject to illness than those who travel daily to and from their work by rail. On the other hand, one of the witnesses told us that, in his opinion, lower house rent at Penrhyndeudraeth, and a better climate, more than made up for the possible risk of catching a chill when travelling. The medical men do not agree with him on this point. The stations and the carriage accommodation of the Narrow Gauge railway might doubtless be improved, and some effort might be made by the railway companies to run more convenient workmen's trains."

The report advised Parliament that at many of the country's ore mines the men had a set of working clothes into which they changed when arriving for work, changing again before returning home in the evening. "A suitable changing-house is provided by the mine owner, with a fire or heating apparatus of some kind for drying the clothes. The miner therefore always has a dry suit in which to commence work, and no matter how wet he gets during the day, he can put on warm, dry and clean clothes before starting for home."

Suggesting the same regime could be beneficially implemented in the slate mines of Blaenau Ffestiniog, the report asked: "But, given the best and most comfortably arranged changing-houses, would the Merionethshire men use them? We must confess that we do not believe they would, and on this ground we do not feel justified in recommending that they should be erected everywhere. We should like to see the experiment tried of providing one of the mines with thoroughly good and comfortable changing-house, with every convenience for washing and dressing; if it were found that the younger men gradually began to avail themselves of the health-giving appliances, we would advise Section 23 (16) of the Metalliferous Mines Act, relating to changing-houses to be enforced rigidly."

They expressed a wish to see the universal provision of comfortable eating-houses, such as were to be found at some of the Blaenau Ffestiniog mines.

"Much was said in evidence about the absence of closet accommodation above and below ground. We are by no means convinced that any illness has ever resulted from this fact; but the present practice may lead to disease, and we recommend that sufficient closet accommodation be provided below and above ground, and that the men be forbidden to use old workings as latrines."

Maintaining the general reluctance to acknowledge the existence and cause of what we now know as silicosis, the report declared: "Considering the large amount of cubic space per man in the slate mills, and the comparatively small amount of dust which is produced, we think that watering with a hose or spray in very dry weather will probably render the atmosphere sufficiently harmless. Every quarryman may, to a great extent, protect himself from the noxious influence of dust by breathing through the nose instead of through the mouth. Men in the mills should be guarded against unnecessary draughts as far as possible."

Under the heading of "general matters," the report recommended: "It is high time that the Metalliferous Mines Regulation Act should be amended. It must be regarded as a tentative Act; it has already been in force more than 21 years, and it has been found in practice to contain various defects. We have pointed out that the average slate mine is more dangerous than the average colliery. If certificated managers are required for the latter, *à fortiori*, they become necessary for the former. In fact any privileges belonging to the collier should be conferred upon the slate-getter. Thus, he should have the power of being represented at inquests, and of making a monthly examination of the workings (Coal Mines Regulation Act, 1887, Section 49, Rule 38). The 21st General Rule of the Coal Mines Act, which compels the securing of the roofs and sides of the travelling

The safety winch devised at Llechwedd by Charles Warren Roberts, for lifting slate blocks underground.

roads and working places, should likewise be made applicable to slate mines. Any new Act should compel a daily visit to every working place by a competent person, and the keeping of a record of any defects or dangers."

Parliament was reminded that interested magistrates were forbidden to adjudicate upon cases under the Coal Mines Act, "and we fail to see any reason why they should be considered sufficiently impartial for dealing with cases under the Metalliferous Act." That was an obvious reference to the proliferation of magistrates among the mine and quarry owners of Blaenau Ffestiniog.

"The present definition of the kind of non-fatal accident which has to be reported gives trouble." Managers and doctors often found it difficult to decide what the consequences of an accident were likely to be, so that some accidents were unreported even though they later proved to be serious. "We consider that the statistics relating to non-fatal accidents would be rendered more reliable by compelling managers to report immediately: (1) any accident causing personal injury which arises from an explosion of gas, any explosive, or any steam boiler; (2) any accidents causing the fracture of a bone of a person being employed, and to report within eight days of its occurrence; (3) any accident which has caused an absence from work of more than seven consecutive days."

It was useless, said the report, to advocate a higher standard of attainments among mine agents and foremen without giving them the means of improving their technical education. "We therefore advise the establishment in Merionethshire of classes for instruction in mining and quarrying, and in the sciences bearing upon these industries. Much useful work of this kind is being done in other parts of the kingdom under the auspices of the respective county councils, and similar work might be done in Merionethshire; we also suggest that the University College of Wales should extend to the mineral industries the aid which they are ready to give to agriculture.

"Further, there is no reason why the time-honoured Eisteddfodau should not direct their attention, now centred almost solely upon literature and music, to hygiene, mining and quarrying. Until the people as a whole are convinced, by precept and by practice, that they can escape many of the ills of the flesh by observing known sanitary rules, we fear that little improvement in the health of the quarrymen can be expected. The well-qualified medical officers of health for the county are quite able to advise what measures should be undertaken; it remains for the ratepayers to support them in their endeavours to promote the welfare of the community.

"As regards Government inspection, one wish of the workmen has been anticipated. Two Welsh-speaking Assistant Inspectors of Mines have already been appointed to the North Wales district," noted the report.

"It will be seen that though some aid from Parliament is invoked, much good may be effected by making better use of existing laws and institutions. It must be remembered, however, as pointed out by Mr. C.Warren Roberts and Mr. Morris Thomas, CC, that the burden of responsibility is divided between the masters and the men. Both have a duty to perform – the former in adopting all possible precautions, the latter in obeying rules and in using such ordinary care as common sense and experience may dictate," concluded the recommendations to Parliament. Their recommendations were quickly given the requisite legislation.

1: John Whitehead Greaves (1807-80) founded the family slate business in 1836, opening up Llechwedd a decade later. 2: His son John Ernest Greaves (1847-1945) ran the business from 1870 to 1945. 3: George Whitehead Greaves (1889-1953), a grandson of the founder, and nephew of J.E.G., was the last chairman to perpetuate the family name, during 1945-53, but he lived in Kenya. 4: Robert Hefin Davies (b.1935) is the present chairman, and was the first non-family member appointed to the board of J.W.Greaves & Sons, in 1971.

LLECHWEDD SLATE MINES

The birth and development of Llechwedd Slate Mines, which figured so prominently in the 1893-94 Parliamentary inquiry, marched hand in hand with the 64-year reign of Queen Victoria. John Whitehead Greaves began seeking slate at Blaenau Ffestiniog in 1836, a year before Victoria's succession to the throne. The Queen died in 1901, two months after J.W.Greaves's children formed themselves into the family limited company that still owns and operates the mines. Three of Queen Victoria's descendants have since visited Llechwedd: David, Prince of Wales; Princess Alice, Countess of Athlone; and Princess Margaret, sister of Queen Elizabeth.

J.W.Greaves was born in Warwickshire in 1807, the third of four sons. His elder brothers were taken into the family banking business from the age of 16, but John was advised to seek his fortune in Canada. He left home in 1830, arriving at Caernarfon, which was then a booming port, and ready source of cheap transatlantic passages with cargoes of slate from the new and busy quayside. He needed no persuasion to put off the misery of a long voyage in the squalid gloom and stench of the accommodation offered to emigrants, and was immediately attracted by the entrepreneurial excitement of Caernarfon's golden age. Quite what he did, or where he lived, for the first three years of his Welsh Odyssey is something of a mystery but he became increasingly involved in the developing slate trade. He met up with another wandering adventurer, Edwin Shelton, wealthy owner of Thorngrove, an estate in the Worcestershire hamlet of Grimley, which he had bought in 1814 from Napoleon's estranged brother Lucien Bonaparte. By 1833 Greaves and Shelton were in partnership, and in 1834 they signed a take-note (short-term exploratory option) on Lord Newborough's Upper Glynrhonwy slate quarry, at Llanberis, rejecting the baron's offer of the nearby Cambrian quarry.

Although Glynrhonwy had been worked since at least 1804, it had just then bankrupted John Robert, of Liverpool, with a debt of £30,000, a colossal sum for that period. Undaunted, the new partners went into the business of supplying Welsh roofs, for the new towns mushrooming around William Blake's "dark satanic mills" of the industrial revolution.

Simultaneously with the collapse of his Llanberis operation, John Robert had lost the lease of Chwarel Llyn Bowydd (Bowydd or "Lord's" quarry) at Blaenau Ffestiniog. This was a surface outcrop quarry opened in 1801, also owned by Lord Newborough, whose brother had worked it during 1823-30, after which it was leased to the hapless John Robert. Lord Newborough lost no time in seeking a new tenant, and in 1834 J.W.Greaves went to stay with Major Edward W. Mathew, JP, at Wern, a house near Porthmadog that was to assume an even more prominent role in the Greaves' family history. They discussed the local belief that John Robert was on the point of success at Blaenau Ffestiniog, after a financially crippling five-year search for good slate, and together they went to look at the derelict Bowydd.

Greaves decided he and Shelton should take it on, and leaving his partner to look after the Llanberis operation, he rented rooms at the Tanybwlch Inn (now the Oakeley Arms), below Plas Tanybwlch, then the home of William Griffith Oakeley. Twenty-one-year leases were signed in 1835 for Glynrhonwy, and for Bowydd together with the adjoining Hafodty Cwmbowydd (usually described as the Foty), and J.W.Greaves rented Tan-yr-allt, a house built by Tremadog's creator, William Alexander Madocks, c.1802. Best-known as the house in which an intruder tried to assassinate revolutionary poet Percy B.Shelley, in 1813, Tan-yr-allt was to remain in the occupation of the Greaves family for the next 150 years. John Ernest Greaves bought it in 1921, for his spinster sister Hilda, and his great-granddaughter, Jean Nagy Livingstone-Learmonth, sold it in 1985.

During 1836 Shelton and his wife moved into Glan William, an elegant house at Maentwrog, after the partners had decided their slate future lay in Blaenau Ffestiniog, rather than their original lease at Llanberis. It is from that final year in the reign of King William IV that the family now count their Blaenau Ffestiniog business anniversaries. It turned out to be an epoch in industry, politics, commerce and social reform. Locally, it was the year in which the Ffestiniog Railway opened its narrow gauge line, to carry the heavy slate wealth of Blaenau Ffestiniog to the ships waiting in the new harbour of Port Madoc, named after W.A.Madocks (and now restyled Porthmadog, in the modern Welsh vernacular). Greaves and Shelton had intended spending £7,000 on laying their own railway for the 8 miles from Bowydd to the old quays beside the River Dwyryd. That was where small boats had traditionally loaded slates, after they had been transported along Casson's badly rutted toll road (from the Wynn Arms to Tanybwlch), first in pack-horse panniers and later in horse-drawn waggons hired at the expensive tariff of 12s (60p) a day. Initially rebelling against the high alternative rail tariff, J.W.Greaves soon accepted its greater convenience, and became treasurer of the Ffestiniog Railway during 1843-47 and 1850-57, and chairman during 1844-48. (His son R.M.Greaves was chairman of FR from 1907 to 1920).

Attracted by the growing wealth of the young Welsh slate ports, the North & South Wales Bank was founded in 1836, financed by the merchants of Liverpool, a maritime city much colonised by the Welsh, and the effective commercial capital of North Wales. Its original branches included Ffestiniog and Porthmadog – and J.W.Greaves became one of its first customers. Much of the commercial history of the area can be gleaned from the subsequent fate of these two branches. Faced with a financial crisis in 1847, the North & South sold its Porthmadog branch to Cassons & Co, a new bank established by the local Diphwys slate quarrying family known to later generations for actor Sir Lewis Casson, and his actress wife Dame Sybil Thorndike, of the Old Vic. The North & South bounced back in 1875, and bought out Cassons & Co. It therefore became reunited with its original 1831 branches, as well as acquiring a Blaenau Ffestiniog branch, in the centre of the new town spawned by the older but more distant village of Ffestiniog, from which it took its name. The North & South was merged in 1908 with the Midland Bank who, in 1986, closed the 150-years-old Ffestiniog village branch, having become but a part-time out-station of Blaenau Ffestiniog. The Midland Bank became part of HSBC in 1998.

In wider social and political terms, 1836 saw the start, in London, of the chartist movement, born out of poverty, hunger and squalor, and seeking remedies in votes for working class men, secret ballots, and other reforms which took several more decades to arrive. It was in that year, too, that the Communist League was born in Paris. It is difficult to guess how much of this kind of news filtered through to the remoteness of Welsh-speaking Blaenau Ffestiniog, but J.W.Greaves appears to have acted out of genuine concern for the well being of his workers when he built ten cottages at Dolgarregddu,

The Victorian management team pose outside the Llechwedd site office. Bowler hats were worn as a badge of rank.

in 1838, at a total cost of £209-7s-11d. These cottages, which are still standing, were let for rents of £3 a year (including 8s ground rent charged by the avaricious Lord Newborough).

By 1840 Edwin Shelton was putting into writing to Lord Newborough his dissatisfaction with the rent and royalty arrangements at Glynrhonwy, which he said made the quarry unprofitable. The correspondence went on for years, and in a letter of March 1843 Shelton added: "No doubt your lordship has heard from Greaves," for his partner had even greater justification for arriving at the same conclusion about Bowydd.

This was at a time when the Great Fire of Hamburg, in 1842, had created an enormous unexpected market for Blaenau Ffestiniog slate, as a new city began to rise from the ashes (an event largely ignored by the Caernarvonshire quarries). Greaves reasoned that an abundance of good slate must lie beneath Llechwedd y Cyd, which was then virgin pasture separating Bowydd and Foty from rival quarries at Rhiwbryfdir. Three landlords, including Oakeley and Lord Newborough, shared grazing rights for 36 cattle on the former Llechwedd common, and negotiation of the three underground mineral leases (in 19, 9 and 8 parts of 36) took until 1846, from which date Greaves abandoned Bowydd (his descendants reabsorbing it in 1975). It was a move which strained the partnership with Shelton, and took Greaves to the brink of financial disaster, as his men used a ton of gunpowder a month, blasting holes, shafts and tunnels all over the site, in a three-year search for good slate. Faltering in his search at Llechwedd, he successfully opened up the neighbouring Maenoffern mine in 1848, but its output was small – its greatest value to Greaves being the geological evidence that the elusive top-quality blue-grey slate must lie hidden somewhere beneath Llechwedd.

Success came in 1849 when his men found the now-famous Merioneth Old Vein, only 50ft below the surface complex at which today's tourists arrive, which is at 849ft above sea level. The foreman raced to Tan-yr-allt to pass on the good news, and although only three years of age at the time, John Ernest Greaves never forgot that fortune-making moment. Once the direction of the slate had been revealed it was relatively simple to locate the lie of the rest of it, in the stratification. Llechwedd was eventually found to be blessed with five slate beds, dipping at an angle of about 30° from the horizontal, and sandwiched between thick layers of hard chert.

Reference has already been made to the comments at the 1893-94 public inquiry about the ingenuity of J.W.Greaves, who revolutionised the slate industry in 1850 with two inventions, his sawing table and his dressing engine. By the end of September 1850 he had joined the big league of slate entrepreneurs with the despatch of 1,128 tons of finished roofing slates.

Having acquired the necessary land at Porthmadog in 1849, and having despatched a few cargoes, he started to build what is still known as Greaves' Wharf, in Porthmadog harbour. It was there, on 23 June 1851, that he opened what was to remain the family office until 1969 – resulting in the anomalous description of his Blaenau Ffestiniog products as Portmadoc slates. Taking stock, in September 1851, after a year of excellent sales, Greaves owned 802 tons of finished slates at Porthmadog, 250 tons at Llechwedd, and 267 tons at Maenofferen. He continued to work Llanberis, and on Caernarfon quayside he had 363 tons, with a further 227 at Glynrhonwy.

Rapidly establishing a reputation as a man of integrity, whose business dealings blended quality with prompt delivery to exclusive dealerships in any given area, Greaves found himself having to fend off would-be buyers, like Jonathan Reynolds of Merthyr Tydfil. "I beg to inform you that I am in the habit of supplying Messrs Watson & Richards, of Cardiff, as also a gentleman in your own town. I am not desirous of extending my connection further in that neighbourhood," he wrote in July 1851.

In October 1851 Greaves won the first of the family's many international prizes: the Class 1 Prize Medal, in the mineral products division at the Great Exhibition, in the purpose-built Crystal Palace at Hyde Park. The citation remains valid to this day, the award being for "strength of material, straightness of cleavage, uniformity of manufacture, and freedom from sulphur or pyrites." Before the month was out he was inundated with inquiries from would-be buyers, like Alfred Ritchie, of Greenwich Wharf, to whom he wrote: "Ever since the success of my slates in the industries department at the Exhibition, in Hyde Park, I have had so great a demand for my slates on the East coast that I have not sought a sale for them in London, but if you will run down by rail to Colchester you may see them in Messrs Hawkins yard."

C.H.Hawkins, of Colchester, was one of Greaves' earliest customers, and during the summer of 1851 they became partners in the building, at Porthmadog, of the schooner

Edith, named after Greaves' eldest daughter (who later became Mrs. Septimus Hansard). This was the romantic origin of what became a floating family tree, taking the name of Greaves, and their slates, across the world. The founder was honoured with the launching, in 1868, of the brig *J.W.Greaves,* and his wife gave her name to the brig *Ellen Greaves,* launched in 1876, the same year as the brigantine *Wern,* commemorating the house where they first met. Their other daughters were similarly honoured: *Constance* (later Lady Henry Smyth) by a schooner, in 1862; *Mabel* (later Mrs. John Clough Williams-Ellis) with a brigantine, in 1864; *Evelyn* (later Lady Osmond Williams) with a brig, in 1877; and *Hilda* (who died a spinster in 1927) with a schooner, in 1877. The heir, John Ernest Greaves, was always too busy to launch a ship, but his wife named the brig *Marianne Greaves* in 1877. The founder's other sons gave their names to the barquentine *Edward Seymour* (1876), and the schooner *Richard Greaves* (1885). The name of J.E.Greaves's 15-years-old daughter (later Mrs.Godfrey Drage) was bestowed on the beautiful three-masted schooner *Dorothy,* in 1891.

Mabel Greaves was the mother of Sir Clough Williams-Ellis (the architect best known for his creation of Port Meirion), of Martyn Williams-Ellis (father of Mrs. Mary Cox, a director of today's J.W.Greaves & Sons), and of Second-Lieutenant Roger Williams-Ellis, who was killed in the Boer War while serving with the 1st Battalion, Royal Welch Fusiliers. Dorothy Greaves was the grandmother of another of today's directors, Mrs. Jean Nagy Livingstone-Learmonth. Evelyn Greaves was the grandmother of Sir Osmond Williams, until recently a director of one of the family subsidiaries (and who as a tank squadron commander in World War Two was awarded the Military Cross in Italy, and the Croix de Guerre and Order of Leopold in Belgium).

Stock returns for 30 September 1852 show how right Greaves had been in pursuing the hidden wealth of Llechwedd. His Porthmadog sheds contained slates valued at £777, nearly all from Llechwedd, where his yards contained finished slates worth another £326. By comparison the Maenofferen stock was valued at only £80, prompting him to surrender the lease before the year was out (although the family reabsorbed the mine in 1975). On Caernarfon quayside his stock was valued at £174, with £98 at Glynrhonwy. (He continued to work Glynrhonwy until 1862, and surrendered the lease on 6 June 1873).

On 3 May 1852 he began building the Floor 2 Mill, now forming part of the tourist complex, in which to start making slate slabs for a variety of purposes, including steps, hearths, window-sills, sanitary installations, billiard tables and industrial chemical plant. He managed to make his first four slabs in August, to open up a whole new market, but it was many months into 1853 before his planing equipment was fully operational.

During all this time Llechwedd was despatching cargoes for the rebuilding of Hamburg. In September 1852 he shipped his first American cargo all the way to San Francisco, then involving the long and hazardous voyage around Cape Horn. His first Australian cargo followed seven months later.

By the autumn of 1852 the fame of Llechwedd was such that Greaves was asked to supply slates for Kensington Palace, birthplace of Queen Victoria. He committed them to the care of Captain Harry Ellis, aboard the Pwllheli sloop *Ann & Ellen,* out of Porthmadog. By a fortuitous coincidence a second consignment was ordered for Kensington Palace in May 1986, while J.W.Greaves & Sons were celebrating their sesquicentenary.

Greaves was at that time obliged to use packhorses or horse-drawn waggons to convey his slates down the steep, rutted and usually muddy hillside to the Ffestiniog Rail-

Here illuminated at the Oakeley mines, by the ingenuity of pioneering underground photographer J.C.Burrow, this study illustrates the perils of working the enormous caverns by candlelight. Until this rare photographic exercise for the Home Office and Parliament, the men had never seen the chambers they had made.

way terminus, for carriage to Porthmadog. In 1853 he completed years of delicate negotiations with rival slate-producing families, who happened to own the narrow strip of land separating Llechwedd from the railway, enabling him to build his incline giving direct access. That marked another surge forward in the fortunes of Greaves. Sales were booming, production was increasing and his order book was bulging. These slates were all extracted from the Old Vein, eventually to be worked down through seven floors beneath the Floor 1 discovery of 1849. In 1867 the miners broke into a new bed of slate, running parallel to, but deeper than the Old Vein, and to which Greaves gave the name of New Vein – although it was geologically older than the Old Vein, hence why it was renamed Deep Vein in 1902. Yet another important bed, the Back Vein, was discovered in 1906.

Following up his success at the 1851 Great Exhibition, J.W.Greaves won an award at the 1862 London Exhibition with slate specimens 10 ft long and 1 ft wide, but only one-sixteenth of an inch thick. At a competition in 1872 everyone marvelled when a Llechwedd quarryman split a block $2^1/_2$ inches thick into 45 sheets. This remarkable property of the original Llechwedd Old Vein is due to the fine grade of its mica crystals, which are only one-2,000[th] of an inch long, and one-6,000[th] in thickness.

John Ernest Greaves, who was born at Tan-yr-allt, came down from Oxford in 1870, just as Plas Weunydd, the mock-Tudor house at the entrance to Llechwedd, was completed by his father, J.W.Greaves. It coincided with the completion of Bericote, another big house he had been building at Leamington Spa, in the area of his youth – from where he had been despatched to seek his fortune forty years earlier. Aged 63, and wealthy, he announced his intention to retire to Bericote, and hand over control of the slate mine to his son, complete with ready-made house on site.

Having been thrust into the chair at the age of 23, J.E.Greaves was destined to control Llechwedd for the next 75 years. He moved into Plas Weunydd and was married in 1875, when his sister Mabel moved out, and two years later married the Reverend John Clough Williams-Ellis, Rector of Gayton, near Blisworth. As the eldest son and heir, in a family of eight brothers and sisters, J.E.Greaves had expected to inherit the family's extensive Avon Side estate of Barford, with the intention of settling down as a Warwickshire farmer. His grandmother Mary had died in 1864, and her Avon Side property passed, via her eldest son, to her grandson Edward Greaves, MP, who had no children. When Edward died, in 1879, J.E.Greaves found he had been left only a walking stick and a canary – his penalty for having produced a daughter (Dorothy) instead of a son. Both Avon Side and the even more extensive Glen Etive estate, in Scotland, passed to J.E.'s younger brother Edward Seymour Greaves, who although not then married was perceived as a prospective father who might eventually produce a son. This Victorian view of a woman's role had a dramatic effect on the future of Llechwedd, for J.W.Greaves promptly changed his will, to give a half-share in the mine to John Ernest, with a quarter each for younger brothers Edward Seymour and Richard Methuen – but nothing for the sisters.

J.W.Greaves died in Blackpool, aged 72, in 1880, the result of a horse-riding accident some four months earlier (he was described in the *Caernarvon & Denbigh Herald* as "one of the most fearless horsemen that ever rode to hounds"). His body was taken back to Bericote, Leamington, for burial at Lillington, and flags were flown at half-mast throughout Porthmadog and Blaenau Ffestiniog. A great pioneering chapter in the Welsh slate industry had ended, but the founder was no doubt delighted with the last returns he received, just a few weeks earlier, showing dressed slates worth £8,908 awaiting shipment from his wharf at Porthmadog, and a reserve stock valued at £5,087 at Llechwedd, with a full order book in the office.

By that time J.E.Greaves had a winter home, Dolfriog, at Nantmor, because the winding 1864 turnpike road we now know as the A470, through the Lledr Valley, was frequently blocked by rain, mud, ice or snow. In 1881 he acquired a third home, Plas Hen, an Elizabethan house at Criccieth. When the Parciau Isaf estate was put on the market in 1883, J.E.Greaves bought the whole of the Bron Eifion land, stretching from Llanystumdwy to Criccieth, and marked out the site for a grand house, designed by himself and his wife. Neither had any architectural training and were never very happy with the end result. The house incorporated a medieval style archway copied from Christ Church cloisters, Oxford, which he had known in his student days, and a chimney canopy copied from a visit to Harlech castle on 10 December 1884. They moved into their new home on 10 October 1885, at the same time putting Dolfriog up for sale, and entering into negotiation with Spanish-born Italian soprano Adelina Patti, Marquise de Caux, then contemplating marriage to French-born tenor Ernest Nicolini. Patti liked the house and the area but eventually chose Craig-y-nos, in the Swansea Valley.

Owning property on both sides of the then county boundary, J.E.Greaves was appointed a Justice of the Peace, High Sheriff and Deputy Lieutenant for Merioneth, in 1884, High Sheriff for Caernarvonshire in 1885, and a Justice of the Peace and first commoner Lord Lieutenant for Caernarvonshire in 1886 (retaining the latter office until 1933). Increasing public work, which began as founder chairman of the Blaenau Ffestiniog local board in 1879, prompted him to hand over the role of general manager at Llechwedd to his brother Richard M.Greaves, in 1885, though retaining for himself the chairmanship of the family syndicate. He served as chairman of Caernarvonshire Quarter Sessions, 1890-1929, chairman of Caernarvonshire County Council, 1906, and Honorary Colonel of the 6th (Caernarvonshire & Anglesey) Battalion, Royal Welch Fusiliars, from its formation in 1908 until 1929. As president of Caernarvonshire T.A.F.A. he had the unenviable task of finding volunteers for the trenches of World War One, and received the CBE in 1918.

R.M.Greaves was married in 1883, and after a lengthy honeymoon in Japan took his bride to Wern, having leased the house from C.A.Huddart. When the Wern estate was auctioned, in September 1886, R.M.Greaves bought it all, and spent the next quarter of a century buying up adjoining land, which became the breeding ground for the prize-winning Wern herd of Welsh Black cattle. Notable Chester architect John Douglas was commissioned to rebuild the house in 1892, and equally eminent landscape specialist T.H.Mawson was engaged to design the gardens.

Anxious to be nearer the Caernarfon courts, councils and committees where he spent so much of his time, J.E.Greaves acquired the Glangwna estate, Caeathro, in 1890, in exchange for Aberglaslyn Hall, another creation of his father, at Nantmor. He later demolished the old house, and having learnt his lesson with Bron Eifion, engaged John Douglas to design the replacement, which was ready for occupation in 1896. Only the library carvings, said to have been inspired by the choir stalls of York Minster, were transferred to the new house. Marianne Greaves designed the panels for the staircase.

On 22 November 1900 the three brothers and their wives (Marianne, Constance Mary and Rosamund Angharad) preserved the name of the founder of Llechwedd Slate Mines when forming themselves into a limited company, J.W.Greaves & Sons Ltd, with John as chairman and Richard as managing director. The workforce was then 608. In 1910 the brothers won a gold medal at the Buenos Aires world fair, to add to those won by their

A busy incline junction at the Oakeley mine. The view below is of a fairly new chamber, being worked by a team of two men who could spend the next two decades extending it inwards and upwards.

father in London in 1851 and 1862, and Paris in 1867. That brought many valuable South American orders for such prestigious buildings as the head office of the National Bank of Argentina.

Edward Seymour Greaves died in 1910, and a year later Rosamund Angharad married Lord Henry Grosvenor, son of the first Duke of Westminster. Lord Henry's nephew, the second Duke, was married in 1901 to Shelagh, daughter of Colonel William Cornwallis-West, of Ruthin Castle. Thus from 1911 until the Duke was divorced in 1919 the former Rosamund Greaves would have had close contact with the notoriously promiscuous Ruthin Castle set, where the colonel's son George was married to Jennie Churchill (mother of Sir Winston Churchill), at a time when George's amorous mother Patsy was dubbed, in a contemporary Sunday newspaper, as the wickedest woman in Wales. The castle's visitors included King Edward VII and his mistress Lilly Langtry, at a time when the king was reputed to be George's real father, and Jennie was known as Lady Randy and one of the king's lovers.

As already noted, Llechwedd Slate Mines pioneered the electrification of industry in North Wales, and with the aim of improving mechanisation the family turned to Martyn Williams-Ellis, the 26-years-old son of Mabel Greaves. He was a highly qualified electrical and mechanical engineer, and agreed to take up the job of general manager and engineer at Llechwedd – but only after he had completed a job in South Africa, estimated to finish in 1914. With the intervention of World War One Llechwedd had to wait until 1918 for the arrival of Captain Williams-Ellis.

His impact was immediate. He upgraded the electrical system and took a critical look at the company's Victorian railway system. In 1921 he bought battery-electric locomotives to replace the horses, and ingeniously converted two steam locomotives to electric power, with overhead trolley pick-up. Every ton of finished slate leaving the mine results in about 15 tons of waste – and the spectacular man-made mountains of rubble that are such a prominent feature of Blaenau Ffestiniog. In 1931 Martyn Williams-Ellis installed an efficient labour-saving aerial ropeway with a J.M.Henderson (Aberdeen) automatic tipping unit. In November 1930 he had revived J.E.Greaves's 1894 idea of removing the overburden, to create an open quarry. By 1931 Williams-Ellis could offer the use of an "American Devil" self-propelled mechanical digger, on caterpillar tracks, to work through the overburden and expose all the underground pillars of top quality slate that had been left behind by the Victorians to support the mines. The family endorsed the plan but had not counted on the greed of three people with inherited unused rights to graze 36 cows on the thin mountain grass that still capped the mine. Three valuable years were wasted in negotiations to buy out these rights, Mrs. Mary Inge eventually netting £20,000 for the Tanybwlch estate's right to graze 19 cows, while Lord Newborough ended his century-old link with the Greaves family when he was paid off with £8,000 for his nine shares.

The cumulative effect of a three-year strike at Penrhyn quarries at the start of the 20th century, and the upheaval of World War One, opened the door for large scale imports of cheap foreign slates. They looked much the same as Welsh slates but were seriously inferior, as highlighted by a case at Liverpool County Court, in January 1931. The plaintiff, William Riley, of 17 Burdett Road, Victoria Park, Waterloo, successfully claimed the cost of re-roofing his 1927 house with Welsh slates instead of the leaking French roof installed by his builder. The judge found the imported slates were "of poor quality, absorbent and liable to crack or chip." Ironically, the defending barrister was the young David Maxwell Fyfe who, after entering politics, became the first Minister for Welsh

Affairs – and was instantly dubbed Dai Bananas.

Following the judgement, Blaenau Ffestiniog Urban District Council convened a conference of local authorities from the Welsh slate areas in August, unsuccessfully appealing for the repeal of Clause 10 of the Housing Act, 1924, requiring materials to be bought at the cheapest possible price for houses built with a Government subsidy. The North Wales Quarrymen's Union reported that 1,300 of their 8,550 members were unemployed, 6,550 were on short time, and only 700 were working full time.

During 1931 the Llechwedd directors were invited to amalgamate with their old rival, the Oakeley quarry, across the road. At face value the approach seemed flattering, for the Oakeley was then employing 832 men, compared with only 416 at Llechwedd. However when consultant engineers and valuers were appointed to examine the proposal in detail, in 1934, they produced some surprising results. Llechwedd was found to be an efficient operation with great potential, and valued at £250,169, which was just double the valuation of £125,631 for the labour-intensive Oakeley, whose remaining underground workings were viewed with suspicion. Wisely, as it turned out, the Llechwedd directors, led by J.E.Greaves, withdrew from the negotiations, and embarked in 1936 on their untopping plan. They lived on independently, to see the once mighty Oakeley close down in 1971. The straggling derelict site was offered to Llechwedd for £7,000 – and refused.

In 1935 Llechwedd was the setting for the first Welsh-language cine film, entitled *Y Chwarelwr* (The Quarryman). It was made for the Urdd cinema, with Sir Ifan ab Owen Edwards as cameraman, and John Ellis Williams as scriptwriter and producer.

J. E. Greaves died in 1945, in his 98th year, his daughter Mrs Dorothy Drage having acted as chairman for the last two years of his life. Anxious to retain the Greaves name, the family appointed Kenya-domiciled Major George Whitehead Greaves, second son of Edward Seymour Greaves, to succeed his uncle in September 1945. On that same day, Lieutenant-Colonel Martyn Williams-Ellis (promoted during World War Two service with the Home Guard) was appointed managing director. His son, John Williams-Ellis, then nearing the end of his war service with the Royal Navy, was also appointed to the board. Martyn Williams-Ellis succeeded to the chair upon George Greaves's mysterious death during the Mau-Mau troubles in 1953. When he died, in 1968, his son, John, succeeded him. At the same time Captain Somerville Travers Alexander (Sandy) Livingstone-Learmonth, was appointed managing director. He was the husband of Dorothy Drage's daughter Cecily, and father of current board member Jean Nagy Livingstone-Learmonth. In 1971 Robert Hefin Davies became the first non-family member of the board of J.W.Greaves & Sons. He became managing director in 1974, and was elected chairman upon the death of John Williams-Ellis in 1989.

In 1947 Mrs. Rachel Williams-Ellis (wife of Martyn) honoured the men of Llechwedd with the gift of a unique two-light stained glass window to mark the centenary of St. David's parish church. One pane shows a rockman, suspended by a chain around one thigh, drilling into a slate face with a pneumatic drill, while a second man works with a crowbar. The other shows a man splitting slate, against the background of a belt-driven Greaves dressing machine.

Photographed at Llechwedd, the view above shows one of C.W.Roberts's "safety bridges". The picture right shows a roof inspection at Oakeley mines, atop an 86-feet high ladder, normally done with no more light than that from the candle in the man's left hand.

Modern tourists returning from an exploration of Victorian slate heritage, aboard the 1846 Miners' Tramway at Llechwedd Slate Caverns, now hauled by battery-electric locomotives.

LLECHWEDD SLATE CAVERNS 12

Llechwedd Slate Caverns (Quarry Tours Ltd), a subsidiary company of J.W.Greaves & Sons, pioneered the opening up of the Victorian Welsh slate heritage to the tourist trade in 1972. Floor 2 Mill, dating from 1852 became the focal point, in which a collection of Victorian machinery was assembled from different parts of the quarry. It was also the starting point for a ride on the old Miners' Tramway into the side of the mountain. Victorian slate wagons provided the chassis for specially designed passenger vehicles, engineered at the Ffestiniog Railway Boston Lodge workshops. Hauled by battery-electric locomotives, the trains pass through some enormous Victorian chambers, in two of which passengers alight for demonstrations of old mining techniques.

The collection of machinery in the Floor 2 Mill includes examples of the inventive genius of the Greaves family. The sawing table by the northern gable wall was J.W.Greaves's original invention of 1850, engineered for him by G.Owen, at the Union Iron Works, Porthmadog. The other table is an 1888 improvement made by the famous De Winton engineering works at Caernarfon. Using an agricultural chaff-cutter as his pattern, J.W.Greaves also designed the first rotary slate-dressing engine – two of the originals are on either side of the entrance to the reception hall. His son Richard patented an improved version in 1886, and this remains the standard equipment at slate works across the world. Two specimens are preserved in the mill, one made by R.Jones & Sons, Porthmadog, and the other by William Lewis, at the Ffestiniog Foundry, Tanygrisiau.

One can only marvel at the dimensions of the 1852 roofing beams, shipped to Porthmadog from Canada. Running the full length of the roof is the old power transmission shaft, and the pulley wheels made by J.Dunnell Garrett, of Suffolk and Ffestiniog. The shaft used to be turned by an enormous overshot water wheel on the site of the adjacent exhibition room. One of the belt drives from the overhead shaft is still attached to all that remains of the original 1890 dynamo. Electricity was used gradually from 1890, and direct waterpower was completely replaced by electricity in 1904, although some steam power continued to be used until 1936. One Robey horizontal engine, at the top of Floor 5's two inclines, was hired out to the Munition Advisory Committee in 1915, and taken to Ffestiniog Railway's Boston Lodge workshops which were converted into a war factory making 18-pounder shells. (R.M.Greaves was a member of the National Shell Factories Board of Management). The engine was not returned after the war. The second Llechwedd Robey steam engine was never re-fired after a long strike beginning in March 1936, when the men demanded a wage increase of two-pence a day.

As the mines got deeper so they gathered water that had to be pumped away, first with the power of surface water wheels and later by electricity. Floors D to I were abandoned in February 1939. All pumping operations stopped on 11 October 1972, when a commemorative plaque was placed at caban Sinc-y-mynydd, on Floor B. Since then the underground workings have been allowed to overflow naturally through an adit on Floor B, which has its exit beside the main incline to the old joint railway sidings of the Ffestiniog Railway and the London & North Western Railway. One of these pumps is preserved in the exhibition room, next-door to the mill.

One of the strangest exhibits is the *car gwyllt* (meaning "wild car"). Hundreds of these were made by the men, using a wheel, a wooden seat and metal parts bought from a black-smith at Manod. The vehicles were used for riding home down the inclines – with frequent accidents. In his annual report for 1898, as regional Inspector of Mines, Dr. Le Neve Foster referred to a fatal accident at the Bwlch-y-Slater slate mine. "One of the workmen lost his life while descending an inclined plane on a so-called 'car.' At the Craigddu quarry, near Blaenau Ffestiniog, there are three inclined planes, having a total length of 1,800 yards, with gradients varying from 8° to 16°. Nearly all the workmen, in order to save themselves the trouble of walking, ride down the inclines upon little vehicles which consist of a low seat supported by a wheel with two flanges, and provided with a small hand-brake. Attached to the frame of the seat is a horizontal bar of iron, 3 ft long with a small guiding wheel at the end. Each inclined plane has two separate lines with a gauge of 23½ inches, which are 2 ft 6 inches apart. The workman places the main wheel of his car upon the inner rail of the road which is on his right hand when descending, and rests the guiding wheel upon the inner rail of the other road. He then sits down upon the seat, and keeping his feet just above the ground, and holding his brake with his right hand, descends the inclined planes under the action of gravity at a very rapid rate. The speed is often so great that the car is known locally as *car gwyllt*. For the average workman, the journey of 1,800 yards, with a [vertical] descent of 1,040 feet, occupies about eight minutes, including the time taken for walking from the bottom of each of the two inclined planes to the top of the one immediately below. When the workman has finished his rapid ride he puts his car into one of the empty slate trucks, and it is drawn up to the top ready for the descent on the morrow.

"For those accustomed to the little vehicle, there is probably no more danger than in cycling or tobogganing; the risk comes when the 'car' is used improperly, that is to say when it is ridden without a regular brake, or is made to carry two persons instead of one. The Bwlch-y-Slater workman on his way home had to pass through the Craigddu quarry, and he borrowed a 'car' to make the descent. It appears that it had no brake. And though some of the practised riders are content to use a piece of stone for the purpose, a novice like the Bwlch-y-Slater workman, who had never ridden on a 'car' before, naturally could not dis-pense with the appliance. His car was upset, he was thrown a long way, and picked up dead."

Dr. Foster added that one of the Blaenau Ffestiniog medical men had recently broken an arm while descending on a *car gwyllt*. This mode of descent was so quick that one could understand why people were tempted to use such locomotion, said the inspector. But the novice should always seek the advice of an old stager, and ensure the 'car' was properly constructed – and refrain from using the vehicle in frosty weather.

In 1979 Llechwedd Slate Caverns added an inclined ride to the tourist experience, using a 24-seat vehicle designed and built by the National Coal Board subsidiary Tredomen Engi-neering. It descends on a Victorian incline with a gradient of approximately 1 in 1.8, and is Britain's steepest passenger railway, into what is now designated the Deep Mine tour. It is a catenary curve, devised for the Blaenau Ffestiniog mines by Charles Spooner, so as to allow for the natural curve taken up by the haulage rope, and therefore reduce abrasive wear on the ground. Victorian miners managed to make this catenary with such precision that the rope actually touches a roof roller when the 3-ton Deep Mine vehicle is at Floor B. The vehicle passes the now sealed Floor 1 entrance to the Old Vein discovered in 1849, and passengers alight another floor down, known as Floor A.

The Victorians designated every level from the initial strike, which became Floor 1. There-after floors going up the mountain were numbered (up to 7), and those going down were lettered (A-I). As part of the Deep Mine tour, visitors walk through a sequence of ten sound

Traditional mining skills are celebrated in stained glass at Blaenau Ffestiniog parish church. This two-light window was the gift of Mrs. Rachel Williams-Ellis, of Plas Weunydd, granddaughter-in-law of J.W.Greaves, to mark the 1942 centenary of the church. Her son John (1923-89) was chairman of the family company from 1968 to 1989.

and light programmes that unfold something of the social history of a typical 1856 slate miner. He ages as one passes from chamber to chamber. Midway through the tour visitors descend to Floor B, where there is a spectacular underground lake that has been used as a film set for two Hollywood productions. The special Deep Mine car, from which they alighted at Floor A, descends to Floor B to collect the visitors at the end of their tour.

Today's tourist complex preserves an interesting collection of Victorian narrow-gauge railway vehicles. Blaenau Ffestiniog's first railway was laid in 1825, within Bowydd quarry. In 1852 J.W.Greaves devised his own rolling stock for Llechwedd, ordering the iron bodies from Caine & Follows, of Shaw's Brow (now William Brown Street), Liverpool, and wheels from local foundries. There was a steam locomotive in use at Llechwedd at least as early as 1879. As already noted, it was 1854 before be could complete his incline to the FR sidings. Using a drum on Floor 3, the twin-track incline was worked by gravity, the descending heavies hauling up the empties.

In 1872 the Cambrian Railways Company brazenly opened exchange sidings at Minffordd, beside the 1836 FR line, thus giving quarries access to the nation's standard gauge rail network instead of having to rely upon ships from Porthmadog. Cambrian immediately captured most of the Llechwedd trade. Seeking the best of both worlds – rail and sea transport – the London & North Western Railway spent five years driving a tunnel 2 miles and 328 yards into Blaenau Ffestiniog, emerging in 1879 at the foot of the Llechwedd incline, while at the same time opening a dock and transhipment wharves at Deganwy, in the Conwy estuary (where conversion into a luxury housing scheme with 50-suite hotel began in 2002). To complete their package the LNWR supplied narrow gauge wagons, specially made for Llechwedd at Earlestown, near Warrington. These could be loaded in the quarry, and sent down the inclines, to be run four abreast onto standard gauge host wagons at a specially made platform at the foot of the Llechwedd incline – where the LNWR metal signs survived until stolen in 1973. Some of this old LNWR rolling stock is still in use at Llechwedd. Yet another bid for a slice of the slate trade was made by the 1882 Bala & Ffestiniog Railway (absorbed by Great Western Railway in 1910), when they completed a $3^1/_2$-mile extension into Blaenau Ffestiniog Central in September 1883 – by the purchase and conversion to standard gauge (4 ft $8^1/_2$ inches) of the 2-ft line of the Ffestiniog & Blaenau Railway Company opened on 30 May 1868. The GWR also supplied narrow gauge wagons, for loading in two parallel rows of three on to host wagons. Specimens of all these 2 ft gauge vehicles, still emblazoned with the names of LNW (and dates 1887 and 1897) and GWR on their cast-iron axle boxes, are displayed at Llechwedd. Elsewhere on site is a 2 ft gauge coal wagon bearing the anachronistic inscription "LMS Earlestown 1898." The London, Midland & Scottish Railway did not come into being until the 1923 network of national combinations, under which the LMS absorbed the LNWR and proceeded to obliterate its name. New axle-box castings were made to replace the name of LNWR, but retaining the date and place of manufacture. Llechwedd also preserves a full range of FR stock; also a Curtis & Harvey gunpowder van – the only private rolling stock permitted to use the FR. These were made out of sheet iron, at the Boston Lodge workshops. There are also wagons peculiar to Llechwedd, with wheels cast at either the Glaslyn foundry of Charles H. Williams, or the Britannia foundry, both at Porthmadog. Also on display is what began as a Bagnall inverted saddle-tank locomotive, No. 1278, when delivered to Llechwedd in August 1890, and named *Edith*. Now unrecognisable as a one-time steam engine, it was converted to electric power in 1931 and renamed *The Coalition*. A matching (though not identical) conversion is to be found in the picnic area, by the upper craft shop. This was originally a Bagnall steam locomotive No. 1568, delivered in September 1899 and given the name *Dorothy*. It was the first to be converted, in June 1927, when it was renamed *The Eclipse*. Each was fitted with a 15 h.p. GEC motor, driven via a trolley from a pair of overhead 230v DC lines. *The Eclipse* remained in use on Floor 5 until 1974. The Ffestiniog Railway's last slate cargo left Blaenau Ffestiniog in August 1946 – with one exception, on 23 October 1994, when Llechwedd despatched a cargo in seven of its wagons, for roofing a new hostel built by FR at Minffordd, for its volunteer workers.

When the slate heritage pioneers began planning Llechwedd Slate Caverns, at the end of 1970, they knew they were establishing Wales's first slate museum, a feature of national importance, but they deliberately eschewed the much misunderstood word "museum" so as to emphasise their broader vision of a living site everyone could enjoy – and thereby absorb the culture with the fun. Since 1972 it has attracted over six-million visitors, including Prince Naruhito (the current Crown Prince of Japan), the Duchess of Gloucester, and the late Princess Margaret. An impressive collection of commemorative slate plaques, at the entrance, tell of modern pilgrimages to Llechwedd by the owners of the salt mines of Austria, the slate mines of Switzerland and the gold mines of Australia.

Never could those Blaenau Ffestiniog miners who gave evidence to the 1893-94 Parliamentary inquiry have envisaged such an audience for the chambers they never saw, beyond the orbit of their candle lights (which remained in use until 1948). Neither could they have imagined that what they created, despite the many problems highlighted by the inquiry, would now be the home of Slate Heritage International. Since opening to the public Llechwedd Slate Caverns have won every major tourism award.

Visitors conversant with the intricacies of heraldry will find a subtle reference to the world's oldest mining heritage in the sign of the Miner's Arms, in the Victorian Village feature at Llechwedd Slate Caverns. The miner whose coat of arms is depicted is R.Hefin Davies, chairman of the company. The shield depicts a falcon which, when linked to the crest showing the Sun disc of the ancient Egyptian gods, flanked by a pair of horns, immediately tells us it is a reference to Hathor – goddess of the underworld, including all the mines of the universe, 5,000 years ago. Altars dedicated to Hathor have been found in the quarters of ancient Egyptian copper miners, in the area now known as King Solomon's Mines, north of modern Eilat, in Israel.

Tourists walking through one of the connecting Victorian tunnels at Llechwedd Slate Caverns. There are 25 miles of such tunnels.

Secretary of State for Wales (and future Conservative Party Leader) William Hague on a visit to Llechwedd in 1995. Below, a painting by June Davis-Miller, showing the original Floor 2 Victorian complex now developed into a multi award-winning heritage site at Llechwedd Slate Caverns.

Above: A modern view from Plas Weunydd. The building on the skyline, top left, is Floor 7 Mill, where roofing slates are made. Below it, and to the right, is the Floor 5 Mill complex, and immediately below that is Floor 3 Mill. The tree in the right foreground conceals the modern tourist heritage site.

Right: The carriage takes today's visitors down Britain's steepest passenger railway, with a gradient of 1:1.8, at Llechwedd Slate Caverns. Its passengers have included Princess Margaret, the Duchess of Gloucester and Crown Prince Naruhito of Japan. Visitors alight for a walking tour through some of the earliest underground workings, now enhanced with ten sound and light programmes that unfold the social life of a Victorian miner.

Above: Victorian miners never saw anything like this! In 1985 a *Penthouse* magazine model shivered her way through the underground workings at Llechwedd, followed by her photographer.

Beauty of another kind enhanced Llechwedd in 1977 when "Miss World" Helen Morgan (left), a friend of the author, turned up to greet the first millionth visitor to the Victorian slate heritage site opened in 1972. She was also the reigning Miss Wales and Miss United Kingdom, but now lives in Spain. By today more than six million visitors have ridden on the Miners' Tramway – seen (right) in Choughs' Cavern.

The gradual revelation, by unfolding lighting sequences, of the huge lake two floors beneath the surface at Llechwedd Slate Caverns is one of the most dramatic episodes of a modern tour. The lake has been used as a set for two Hollywood films (*Prince Valiant* and *Black Cauldron*) and its shoreline has become a popular wedding venue.

The picture (left) shows modern machinery working through the slate pillars that were left behind by the Victorian miners before the uncapping of what is now an open quarry.

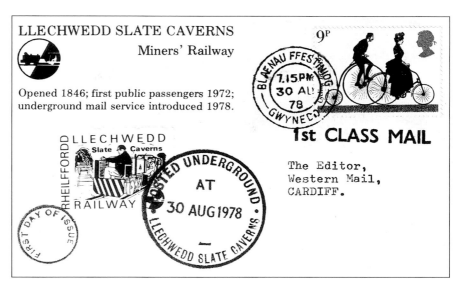

LLECHWEDD SLATE CAVERNS

Miners' Railway

Opened 1846; first public passengers 1972; underground mail service introduced 1978.

LLECHWEDD
Slate Caverns
RHEILFFORDD
RAILWAY

POSTED UNDERGROUND
AT
30 AUG 1978
LLECHWEDD SLATE CAVERNS

FIRST DAY OF ISSUE

BLAENAU FFESTINIOG
7.15PM
30 AU
78
GWYNEDD

9P

1st CLASS MAIL

The Editor,
Western Mail,
CARDIFF.

Above: Before the centralisation of all North Wales postmarking at Chester, the underground railway letter post, using Victorian legislation, was a popular feature at Llechwedd. It created what are now highly prized collectors' items – one of which is in the Royal Philatelic Collection. Below: Among the features in the exhibition hall at Llechwedd Slate Caverns is the model of one of the many Victorian water wheels that used to power the site. The exhibition was opened by the grandson of author Dylan Thomas, during a school visit in1986.

OPENED BY
HUW DYLAN ELLIS
GRANDSON OF DYLAN THOMAS
13TH JUNE 1986

VICTORIAN
SLATE MINING
IN PHOTOGRAPHS

MILLS

An impressive assembly of the slate sawing tables invented by J.W.Greaves, in one of the enormous production mills at Blaenau Ffestiniog.

Horses were used for haulage at Blaenau Ffestiniog until the 1920s.

One can but marvel at the length and strength of Canadian beams imported at Porthmadog, and hauled up the unsurfaced roads to Blaenau Ffestiniog, for roofing the slate mills shown in these 1894 photographs, taken by J.C.Burrow at Oakeley (above) and by C.Warren Roberts at Llechwedd (below).

SURFACE

Floor 5 Mill at Llechwedd, showing both a working water wheel and a steam-driven engine house. There is another water wheel on the hill above. The picture opposite top shows both a water wheel and steam engine-house used for incline haulage. The picture opposite bottom shows a Ffestiniog Railway brake truck.

Busy 1894 production scenes at Llechwedd Slate Mines, Floor 5 above, and floors 2 and 3 below. Today's visitors will be able to identify the Floor 2 buildings, most of which have been preserved in the heritage site (see also page 90).

A steam-powered incline from Floor 5 at Llechwedd (above). Part of the Foty & Bowydd quarry (below), photographed in 1894 by G.J.Williams, to show the opencast exploitation of supporting pillars from earlier mining activity. Foty & Bowydd now forms part of the Llechwedd complex.

Battery-electric locomotives (made by BEV) were introduced to Blaenau Ffestiniog in the 1920s to replace horses. The block-carrying wagons, seen right, were designed by Richard M.Greaves, in both three and four-bar versions.

Illegal man-riding on a Blaenau Ffestiniog incline, showing a rubble wagon and a block-carrier. The photograph below is taken from the site of the old Llechwedd barracks, high above Plas Weunydd, to show the Talwaenydd community on the right, against a backdrop of the Oakeley quarries.

A Bagnall steam locomotive delivered to Llechwedd in 1890 can be seen in the above photograph, which also shows a water-balance incline in use. Slate was raised from below aboard a wheeled water tank, from which the water was gradually released to become counter-balanced by a full water wagon descending from above. One of the many Llechwedd water wheels can be seen on the skyline of the picture below.

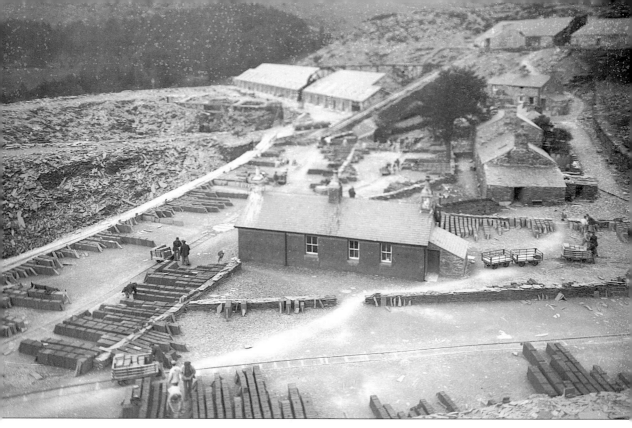

An 1894 view of Floor 2, at Llechwedd, now readily identifiable as the Victorian heritage site seen by tourists. The photograph below shows the aerial waste removal system installed by Martyn Williams-Ellis, a grandson of J.W.Greaves.

Manual slate tipping, to help build up the mountains of waste that encircle Blaenau Ffestiniog. The photograph below was taken by J.C.Burrow at the Oakeley quarries, showing the remains of earlier underground chambers.

Slate stacking yards at Llechwedd Slate Mines. The little building near the right edge of the top picture was a row of pioneering self-flushing water closets, built over a stream which carried the problem away to someone else's territory! They may now be viewed by tourists.

This view in the open quarry at Llechwedd was used as a Victorian postcard.

PORTHMADOG

An early scene (above), seemingly in the 1880s, at Porthmadog harbour, showing the Oakeley wharves on the left and in the foreground, and vessels tied up at the Greaves wharf on the right, with the empty Holland wharf beyond. Continuing from the bottom right, beyond the view of the camera, were the wharves for Diphwys Casson, Foty (F.S.Percival), Maenofferen, Cwm Orthin and Ffestiniog Slate Company. The Welsh Slate Company's wharf was top left, commencing at the Britannia Bridge, and was absorbed by Oakeley, as was the Holland wharf. All the wharves were served by Ffestiniog Railway, directly from the mines and quarries in Blaenau Ffestiniog.

The 1894 view, as seen by J.C.Burrow from the opposite direction, with Britannia Bridge in the foreground, is shown in the top photograph opposite. The Greaves wharf is seen in the bottom photograph, with three types of Ffestiniog Railway slate wagon in the foreground. That on the left is one of the very earliest locally made wooden wagons, No.107. That in the middle is one of the small iron wagons introduced in1857. The larger iron wagon, on the right, was first intro-duced in 1869 but was unpopular with the loaders and not many were manufactured.

Well-stocked slate wharves at Porthmadog in 1894. The bottom picture shows slates being slid in loose bundles down a wooden plank for packing into the schooner's hold, where they were wedged into place with straw. It was these wharves that resulted in Blaenau Ffestiniog's products being marketed, world-wide, as "Best Portmadoc slates" (using the original spelling of the town's name).

Porthmadog's famous wooden paddle-tug, the *Wave of Life,* is seen moored at Greaves's wharf, alongside the two-masted schooner *Edith Eleanor.* Behind it is the brigantine *C.E.Spooner.* They were photographed by J.C.Burrow. The photograph below shows the immaculate interior of the Greaves storage sheds, on the quayside.

ROLLING STOCK

A BEV-made battery-electric locomotive delivered to Llechwedd Slate Mines in 1921.

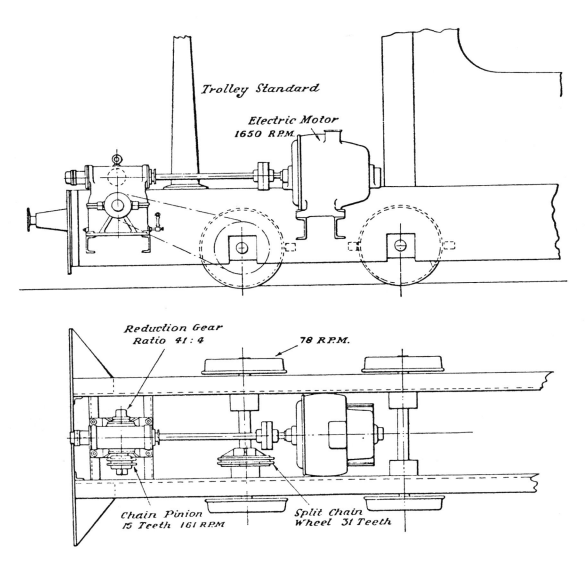

Trolley Standard

Electric Motor
1650 R.P.M.

Reduction Gear
Ratio 41 : 4

78 R.P.M.

Chain Pinion
15 Teeth 161 R.P.M.

Split Chain
Wheel 31 Teeth

Martyn Williams-Ellis's drawings for his novel conversion of Llechwedd steam locomotives to electric power, using an overhead trolley pickup.

Bagnall inverted saddle tank steam locomotive No.1278 (above) was delivered to Llechwedd in August 1890, and named *Edith*, after a daughter of J.W.Greaves. In 1931 it was converted into the electric locomotive *Coalition,* below, now preserved at Llechwedd Slate Caverns.

Bagnall saddle tank steam locomotive No.1568 (above) was delivered to Llechwedd in September 1899, and named *Dorothy*, after another daughter of J.W.Greaves. In 1927 it was converted into the electric locomotive *Eclipse*, below.

Some of the rare Great Western Railway narrow gauge slate wagons delivered to the mines and quarries of Blaenau Ffestiniog. These, and their cargoes, were assembled on standard gauge wagons, specially fitted with rails, for transportation on the English main line network. The above wagons are on display at Llechwedd, as are the various Llechwedd Railway wagons shown below.

The London & North Western Railway also supplied narrow gauge slate wagons (below), for loading onto their standard gauge trucks, for the 30-mile journey to the railway company's own purpose-built dock at Deganwy, on the Conwy Estuary (now in the process of being developed into a luxury housing estate).

The rheostat (left), is all that remains of the controls in the Williams-Ellis conversion of Llechwedd steam locomotives to electric power. The vehicle above, photographed at Llechwedd, is a Curtis & Harvey gunpowder wagon; the only privately owned rolling stock allowed on the Ffestiniog Railway. The strange vehicle below was used for rail-bending.

RAILWAYS

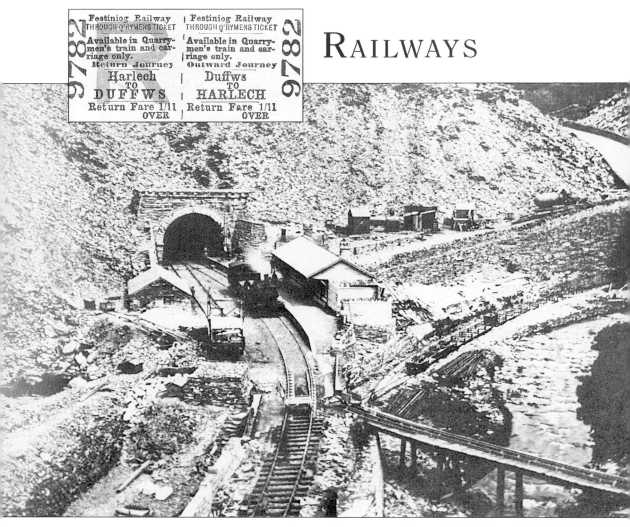

The splendid photograph above shows the first Blaenau Ffestiniog railway station, opened on 22 July 1879, at the mouth of the two-mile-long LNWR tunnel. The narrow gauge train on the right of the picture is made up of empty wagons waiting to be hauled back to Llechwedd, on the funicular incline. The full slate wagons were shunted onto the bridge in the foreground, from where they were loaded onto standard gauge trucks.

The illustrations opposite are of the rival eastbound railway. The upper picture shows narrow gauge slate wagons being unloaded off a standard gauge GWR truck at Blaenau Ffestiniog Central station. The bottom picture, by John Thomas, shows a train of the Bala & Ffestiniog railway leaving Blaenau Ffestiniog in 1883.

Blaenau Ffestiniog North station (above) was the eventual terminus of the LNWR, but this has now closed, with the insertion of an extra length of track into Blaenau Ffestiniog Central (below), the severed rump of the old GWR line (which was closed in 1960 by the creation of Llyn Celyn reservoir). The new terminus also serves the Ffestiniog Railway.

Llechwedd's first steam locomotive was a De Winton vertical boiler "coffee pot" made at Caernarfon, similar to the one above. The fire was fed through a trapdoor and chute in the footplate. Two vertical cylinders operated a direct drive to the front axle. The picture below was taken on the Ffestiniog Railway, at Tanybwlch station, and shows a combination of slate wagons and the workmen's carriages condemned as a health hazard by the 1893-94 Home Office Departmental Committee.

𝔓𝔥𝔬𝔱𝔬𝔤𝔯𝔞𝔭𝔥𝔦𝔠 𝔖𝔬𝔲𝔳𝔢𝔫𝔦𝔯

AND MAP in connection with the

LAST PASSENGER TRAIN

on the

BALA-BLAENAU FESTINIOG BRANCH
of former Great Western Railway

SUNDAY, 22nd JANUARY, 1961

Organised by

THE STEPHENSON LOCOMOTIVE SOCIETY
(Midland Area)

Chronology of the line

1868, 30th May. Festiniog and Blaenau Railway (n.g.—1′ 11½″) opened from Festiniog—Duffws, 3½ mls., where it met the Festiniog Railway.

1882, 1st November. Bala and Festiniog Railway opened, 22 mls., and worked by the Great Western Railway.

1883, 13th April. Festiniog and Blaenau Railway vested jointly in the B. and F.R. and the G.W.R.

1883, 10th September. Conversion of the F. and B.R. to standard gauge completed, permitting through running from Bala—Blaenau Festiniog.

1910, 1st July. The B. and F.R. and the F. and B.R. acquired by the G.W.R.

1960, 2nd January. Last day of passenger service.

1961, 27th January. Last day of goods service.

The valley between Arenig and Frongoch is scheduled for building of a Liverpool Corporation Reservoir and will involve the flooding of the sites of Capel Celyn and Tyddyn Bridge Halts.

As most of the line runs through mountainous country it is very steeply graded, very largely at 1—50 to 1—75. It rises from Bala to Cwm Prysor, falls to Maentwrog Road and rises again to Manod.

EQUIPMENT

A demonstration for the 1893-94 Departmental Committee of the safety winch devised by Charles Warren Roberts for underground derricks used to lift newly quarried blocks.

Slate mining pioneer John Whitehead Greaves invented several machines for the industry but the only one for which he registered a patent was the slate dressing engine, shown on these pages. Originally they were hand-driven, like the one above, but were soon adapted for belt-drive from an overhead shaft driven by a water wheel. With embellishments for a new patent taken out by his son Richard M.Greaves, the Greaves dressing engine remains in use as standard equipment at all the world's slate mills. Today's machines are turned by electricity.

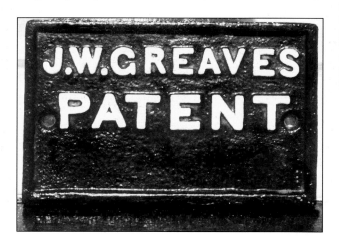

Above, detail of C.Warren Roberts's safety winch. Still preserved at Llechwedd Slate Caverns is one of the slate cutting knives (below) used before the invention of the Greaves dressing engine.

Modern visitors (above) to the slate heritage mill at Llechwedd examine the sawing table invented by J.W.Greaves. Below are the remains of the self-powered funicular winch on Floor 3 at Llechwedd, once used to lower finished slates, in their narrow gauge wagons, to the joint Ffestiniog Railway and LNWR/LMS/British Rail standard gauge railhead at the foot of the quarry's main incline.

A demonstration to Blaenau Ffestiniog mine managers of the Doering & Sachs pneumatic boring machine, also shown in the drawing opposite. It was much more powerful than the drill shown below, which was advanced by a hand-crank from a fixed frame.

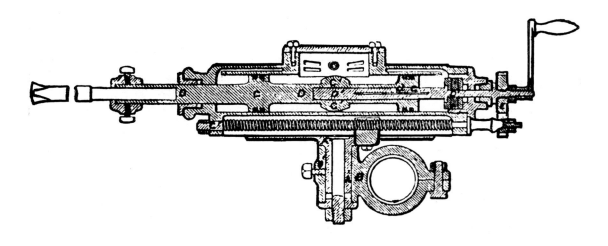

The bottom photograph, is of a *car gwyllt* ("wild car") man rider, used by quarrymen at the end of the day, so as to leave work quickly by riding down the inclines, resulting in many serious injuries and deaths.

The remains of an early electric incline winch at Llechwedd (above). The electrically driven pump (below) was used underground from 1910 until 1970.

APPENDIX 1
19TH CENTURY JAPANESE LINK

The Victorian slate miners of Llechwedd had a little-known naval link with Japan, now celebrated with an interesting plaque at the reception hall for modern tourists. Inscribed in Japanese and English, and displaying the Imperial Chrysanthemum, it commemorates the 1985 visit of Crown Prince Naruhito, in celebration of an event a century earlier when general manager Richard M. Greaves was appointed consultant engineer to the embryo Japanese Navy. Two Japanese ambassadors have since followed in the footsteps of the Prince, His Excellency Hiroshi Kitamura, in 1992, and His Excellency Sadayuki Hayashi, in 1998.

The plaque is also a subliminal reminder of Llechwedd's tenuous link with Nagasaki's best known citizen, Chô-san, elevated into eternal theatrical orbit by Giacomo Puccini in his opera *Madama Butterfly,* after the composer's 1900 visit to London's Duke of York Theatre, to see a play of the same title dramatised by David Belasco from a story by John Luther Long.

Richard M.Greaves was born in 1853 at Tan-yr-allt, Tremadog. As already noted, he joined De Winton & Co, who were then building both industrial railway and marine steam engines at Caernarfon. As a mark of good faith in his work, he signed on as 3rd engineer for the 1872 maiden voyage from Liverpool to South America of the SS *Mimosa* (541 tons) fitted with a De Winton engine. He was married in 1883 to Constance Dugdale, of Wroxall Abbey, Warwickshire, and joined the family business at Llechwedd in 1885.

It was during their 1883 honeymoon that Greaves and his bride found themselves among the earliest tourists to Nagasaki, then a port of some 50,000 inhabitants, including several hundred foreigners who had been described a couple of decades earlier as "Californian adventurers, Portuguese desperadoes, runaway sailors, piratical outlaws and the moral refuse of the European nations." As the first Japanese harbour to establish trading contacts with the outside world, Nagasaki had also attracted many Christian missionaries, albeit outnumbered by barmen and prostitutes.

Richard Greaves's skills as a marine engineer became known within the social chatter of this foreign community and he was approached at his hotel by officers of the US Navy, who told him they had a problem with one of their three warships anchored in the harbour. We cannot be sure of the identity of the vessel, but Greaves appears to have had some contact with the USS *Monocacy,* a side-wheel gunboat which was in Nagasaki during July and August 1883, and again in October and November, but a search of her log books at the National Archives, in Washington, DC, gives no hint of any engineering problem during that year – the *Monocacy* served with distinction in the Asiatic Station from 1864 until 1904, and Greaves may have been entertained on board merely as a guest.

The most likely candidate for Greaves's engineering skills was the USS *Palos,* which was also frequently in and out of Nagasaki that year, as part of the Asiatic Station fleet. She was constantly breaking down and eventually had to be towed into Nagasaki in 1892 for decommissioning. We shall return later to the *Palos.*

Greaves was able to help the US Navy put to sea, and as they sailed away he was approached by an emissary for the young Japanese Navy, who had been concealing an engineering problem aboard one of their own warships, the *Hiei,* a 2,248-ton corvette. While inspecting the *Hiei* Greaves was surprised to learn from her maker's plates that she was built in Wales, at the Pennar shipyard of the Milford Haven Shipbuilding and Engineering Company, at Pembroke Dock, founded by naval architect and politician Sir Edward Reed, under the chairmanship of Lord Clarence Paget, son of the first Marquess of Anglesey. Designed by Reed, the *Hiei* was the first major vessel to be launched at Pennar, on 12 June 1877. The event was preceded on 8 June by a splendid dinner party, hosted by Reed at his London home. Guests included the ambassadors of Japan (Jushie Wooyeno Kagenori) and China, Baron Reuter, founder of the famous international news agency, and Heinrich Schliemann, the controversial archaeologist of Troy fame. Guests at the subsequent Pembroke launch included Prince Hashisuke, one of the wealthiest of the pre-imperial feudal daimyos of Japan.

The corvette was completed nine months later, on 23 March 1878. With portholes on either side, the *Hiei* was equipped with three 6.7-inch Krupp breach-loading guns, six 5.9-inch Krupp guns, four 1-pdr guns, seven machine-guns and two 14-inch torpedo tubes.

Greaves was able to remedy the problem on the *Hiei* and he returned to Wales with the appointment of consultant engineer to the Imperial Japanese Navy. Thereafter he was a frequent visitor to Japan and was involved in the wars against China in 1894 and Russia in 1904. The 1902 autograph of Richard M.Greaves is still displayed in the visitors' book at an antique shop in the town of Nikko, together with that of Captain Henry Cadogan, of the Royal Welch Fusiliers. They had teamed up in Japan while Cadogan was on his way home to North Wales from Peking, where he had stayed on with the allied mounted infantry, after the suppression of the 1900 Boxer Rebellion against foreign nationals – in which Japanese troops fought alongside the 2nd Battalion of the Royal Welch Fusiliers.

Cadogan was killed while serving in Flanders during World War One, as Lieutenant-Colonel commanding the 1st Battalion, Royal Welch Fusiliers. Greaves died in 1942, dismayed that Britain and Japan were locked in a particularly cruel war. The old engineer could never have imagined that the bitter war would be brought to an end with the nuclear fission technology which destroyed his much-loved Nagasaki.

Britain and Japan, who had fought as allies in World War One, might never have gone to war in 1941 if one of R.M.Greaves's county contemporaries, barrister John Robert Jones, of Llanuwchlyn, had not taken such a slow boat from China in 1940. J.R.Jones had been a leading figure in the Shanghai Supreme Court since 1924, and in 1928 became Secretary General of the Council at the International Settlement, a semi-autonomous community of 85,000 people of 49 different nationalities (mostly Japanese). At different times he served as secretary and president of the Shanghai branch of the Royal Asiatic Society.

Earlier in his life J.R.Jones had been involved in many strange missions. He wrote the three standard books on the obscure Etruscan language of Northern Italy, which had no links with the European family of languages. While exploring the ruins of Carthage, in 1911, he was asked by fellow Welshman David Lloyd George to do some spying into the political problems between the Riffs and the French. He joined the local defence unit of the French Foreign Legion at Casablanca and arranged to be captured by the Riffs, subsequently supplying Lloyd George with a detailed report.

When Lloyd George sent him to Ireland in 1914 to recruit a battalion for the war in Flanders, J.R.Jones raised a whole brigade and served in France and Flanders for the rest of the war, earning the Military Cross for gallantry. He was serving with the British army of occupation in the Rhineland when Lloyd George next called upon him to do some spying in the Ukraine, as liaison officer with the nationalist forces opposed to the Bolshevik Revolution.

It came as no surprise to J.R.Jones when he was summoned to Tokyo by Emperor Hirohito, who entrusted him with a secret message inviting Britain to restore the old Anglo-Japanese alliance, which London had terminated in the 1930s. Once more he donned the cloak of Jones-the-Spy and set off aboard a ship that docked several weeks later in neutral Italy. It was June 1940, and with the Germans already in Paris, he rushed to Rome to place the Emperor's message in the diplomatic bag at the British Embassy. Before the day was out Benito Mussolini sent his troops into Southern France and declared war on Britain. The French government surrendered within 48 hours, and the diplomatic bag from Rome never arrived in London. Three months later Japan's military-controlled government signed the Axis pact, claiming German and Italian supremacy in Europe while giving the Japanese supremacy in the Pacific. When Japan entered the war at the end of 1941 J.R.Jones was trapped in Shanghai, but was given special status and sent to Canada by neutral ship. He died in Hong Kong in 1976.

What of the USS *Palos*, the prime candidate for the warship R.M.Greaves helped to repair at Nagasaki in 1883? We know the libretto written by Giuseppe Giacosa and Luigi Illica for Puccini's 1904 production of *Madama Butterfly* was based upon an 1898 short story, which emerged as a stage play in 1900. We also know, from the writings of his Methodist missionary sister Jennie Correll, that John Luther Long's story was based upon actual events while she was in Nagasaki.

Seemingly there was an aristocratic Samurai called Chô-san, working in a teahouse where she was known as Butterfly, and who, at some time between 1892 and 1894 was reduced to the indignity of being "married" into temporary concubinage with an American naval officer - the Lieutenant Benjamin Franklin Pinkerton of Puccini's opera.

Pinkerton can now be identified as Ensign William Benjamin Franklin, born in January 1868, who served at Nagasaki several times between 1892 and 1894. He was aboard the USS *Marion* in 1892 when she took the troublesome *Palos* in tow at Shanghai, and delivered her into a dry dock at Nagasaki. Aboard the *Palos* was Naval surgeon John Sayre, who was actually named in John Luther Long's *Madame Butterfly,* and Sayre and Franklin are known to have been shipmates at Nagasaki for eight months.

Naval records at Washington show that while at Nagasaki, on 5 July 1892, Ensign Franklin reported sick with gonorrhoea, presumably a wedding present from the virginal Butterfly of Puccini's opera. Sayre also seems to have entered into a "Japanese marriage," a brokered arrangement for the convenience of visiting officers. He was discharged from the Navy with tertiary syphilis in 1896.

Tellers of the original Nagasaki story made no mention of Chô-san's having ended her life on her father's sword - seemingly a fictional postscript designed to ensure the immortality of Puccini's Butterfly.

APPENDIX 2

ALICE'S SHOP

Alice's Shop, in the reception hall at Llechwedd Slate Caverns, is reminiscent of the window of the famous Victorian tea shop illustrated by John Tenniel in Lewis Carroll's *Through the Looking Glass*. It was a shop the real-life Alice Liddell knew well during her childhood and youthful years across the road, at Christ Church Deanery, Oxford.

Alice Liddell, for whom Lewis Carroll wrote his immortal books *Alice's Adventures in Wonderland* (originally called *Alice's Adventures Under Ground*) and *Through the Looking-Glass and What Alice Found There*, was seven when she spent her first holiday in Wales, at Penmaenmawr, in 1859. Although probably sleeping at Plas Mariandir, they were holiday guests of Manchester solicitor Samuel Darbishire who, in 1858, had moved his family into Pendyffryn, a handsome 20-bedroom Georgian house which had a drive to its own railway halt on the Chester-Holyhead railway line.

Samuel Darbishire was a friend and life-long host of politician William Ewart Gladstone and physician [Sir] Henry Acland. In 1859 Gladstone, whose home was at Hawarden Castle, in Flintshire, was MP for Oxford University and Chancellor of the Exchequer. Acland was teaching medicine at Oxford University, and acted as personal medical adviser to the Dean of Christ Church, the Very Reverend Henry George Liddell, and his family.

It was obviously on the recommendation of his friends Gladstone and Acland that Dean Liddell was able to take his family to Pendyffryn for a long holiday in 1859, and in all probability in 1860. At Pendyffryn Alice Liddell met one of Samuel Darbishire's granddaughters, Marianne Rigby, aged 8, (always known within the family as Polly). She and her younger sister Susie were the daughters of Dr Edward Rigby and Marianne, senior (née Darbishire). Mrs. Rigby had died six years earlier, aged 25. Their father, an obstetrician practising in Berkeley Square, London, was in the habit of sending the young sisters to their grandparents Samuel and Mary Darbishire for long summer holidays. Tragically he, too, was to die young in 1860, after which the orphans made their home at Pendyffryn.

In 1875 Marianne Rigby married John Ernest Greaves, son of the founder of Llechwedd Slate Mines. Alice Liddell was married a year later, to Reginald Hargreaves, of Cuffnells, Emery Down, in the New Forest. Both women died in 1934.

Upon their marriage John and Marianne Greaves moved into Plas Weunydd, the large mock-Tudor house at the entrance to the mines. Marianne took with her a cherished copy of *Alice in Wonderland*, published in 1865, only six years after she and the real life Alice had first played together in the ample grounds of Pendyffryn, and had worshipped together at the little parish church of Dwygyfylchi (since replaced by the present St.Gwynin church).

By a strange twist of fate, Louise Darbishire, sister of Marianne, senior, was the grandmother of Somerville Travers Alexander (Sandy) Livingstone-Learmonth who, in 1931, married Cecily Drage, granddaughter of Marianne Greaves. Sandy Livingstone-Learmonth was destined to become managing director of J.W.Greaves & Sons, at Llechwedd, in 1968.

From their temporary holiday home at Penmaenmawr the Liddells made their first visit to Llandudno, on the opposite shore of the Conwy Estuary, where the Dean was much impressed by the new resort that had started to sprout on the ancient marsh and sand dunes forming the isthmus linking the Great Orme to the mainland. Using his Penmaenmawr contacts, the Dean arranged to take his family to Llandudno for the Easter of 1861, to stay at Tudno Villa, at the northern end of the promenade.

On his way to church, during that Easter holiday, the Dean spotted a drawing for a proposed house in an architect's window. It was for a strangely alien Gothic-inspired monstrosity but the Dean liked it, commissioned the plans and leased a choice plot of land at what was then called Penmorfa, the descriptive Welsh name (i.e. End of the Sea Marsh) for the remote rabbit infested sand dunes on Llandudno's West Shore, overlooking Conwy Estuary and the Menai Strait. The house was completed in August 1862 and remained the Dean's property until sold in 1873, retaining the name of Penmorfa.

Letters written at Penmorfa tell us the house was used regularly for long Easter, Summer and Christmas vacations, and the Liddells hosted many distinguished visitors to Llandudno. Although the Dean would return to Oxford earlier, his family seemed to enjoy very long Christmas holidays, as we see from a letter which Alice wrote to her father from Penmorfa on 5 February 1863, for his birthday the next day: "Yesterday we had another gale. Edith [aged 8] and I [10] had our bed moved down into Harry's [16] room, we lay on two mattresses on the floor and were very comfortable indeed. Little Rhoda [3] had her little cot down into Mama's room, where Pickey [i.e. governess Mary Prickett] and Ina [13] slept, and Mama was in the Blue Room, altogether we were very jolly."

From Penmorfa the family travelled extensively to the big houses of the region, ranging from Penrhyn Castle in the west, to Kinmel in the east, and they maintained their links with the Darbishires and the young Marianne Rigby at Penmaenmawr, a place they could see from their living room, across the Conwy estuary.

Alice's cousin Florentia Hughes (née Liddell) had lived since 1853 at Kinmel, the mansion of H.R.Hughes hidden in its splendid parkland beside the present A55 road near Bodelwyddan. Alice and Florentia were great-granddaughters of Sir Henry George Liddell, the 5th baronet.

Marianne Greaves cherished her links with Alice, now celebrated in Alice's Shop. In recognition of its tenuous family link with the childhood Alice and her underground wonderland, Llechwedd issued a se-tenant pair (Welsh and English) of special stamps for use on its underground railway letter post, on 18 July 1979. The stamps reproduced a slightly amended version of one of the original Tenniel drawings, in which Alice watches the White Rabbit disappear down a black hole - in the Llechwedd version Alice is holding a miner's hammer, and has a chisel at her knees. Another Tenniel drawing, showing Alice on a train journey, featured on a later Llechwedd stamp.

FURTHER READING

Davies, D.C.: *Slate and slate quarrying,* London, 1878.

Drage, Dorothy: *Pennies for Friendship,* Criccieth, 1961.

Ffestinfab: *Hanes Plwyf Ffestiniog,* Blaenau Ffestiniog, 1879.

Hatherill, Gordon & Ann, *Slate Quarry Album,* Garndolbenmaen, 2001.

Holland, Samuel: *The Memoirs of Samuel Holland,* Dolgellau, 1952.

Jones, Ernest: *Senedd Stiniog,* Bala, 1975.

Jones, Ivor Wynne: *Eagles do not catch flies,* Blaenau Ffestiniog, 1986.

Jones, Ivor Wynne: *Gold, Frankenstein and manure,* Blaenau Ffestiniog, 1997.

Jones, Ivor Wynne: *Llechwedd, and other Ffestiniog railways,* Blaenau Ffestiniog, 1977.

Jones, Ivor Wynne: *Minstrels and Miners,* Blaenau Ffestiniog, 1986.

Jones, Ivor Wynne: *Shipwrecks of North Wales,* Ashbourne, 2001.

Jones, Ivor Wynne: *The Llechwedd strike of 1893,* Blaenau Ffestiniog, 1993.

Jones, R.Merfyn: *The North Wales Quarrymen, 1874-1922,* Cardiff, 1981.

Lewis, E.Llewelyn: *The Slate Industry,* Denver, USA, 1927.

Lewis, M.J.T. & Williams, M.C.: *Chwarelwyr cyntaf Ffestiniog,* Tan y Bwlch, 1987.

Merionethshire Slate Mines, Report of the Departmental Committee, HMSO, 1895.

Roberts, Dafydd G.: Moses Kellow a chwareli Cwm Croesor, *(Cylchgrawn Cymdeithas Hanes a Chofnodion Sir Feirionnydd, Cyfrol IX),* 1982.

Williams, G.J.: *Hanes Plwyf Ffestiniog,* Wrexham, 1882.

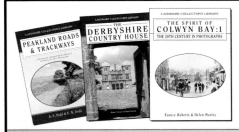

LANDMARK
COLLECTOR'S LIBRARY

LANDMARK
Publishing Ltd ↑••••

Ashbourne Hall, Cokayne Ave, Ashbourne, Derbyshire, DE6 1EJ England
Tel 01335 347349 Fax 01335 347303
e-mail landmark@clara.net web site: www.landmarkpublishing.co.uk

Mining Histories

- Collieries of South Wales: Vol 1   *ISBN: 1 84306 015 9, £22.50*
- Collieries of South Wales: Vol 2   *ISBN: 1 84306 017 5, £19.95*
- Collieries of Somerset & Bristol   *ISBN: 1 84306 029 9, £14.95*
- Copper & Lead Mines around the Manifold Valley, North Staffordshire   *ISBN: 1 901522 77 6, £19.95*
- Images of Cornish Tin   *ISBN: 1 84306 020 5, £29.95*
- Lathkill Dale, Derbyshire, its Mines and Miners   *ISBN: 1 901522 80 6, £8.00*
- Rocks & Scenery the Peak District   *ISBN: 1 84306 026 4, paperback, £7.95*
- Swaledale, its Mines & Smelt Mills   *ISBN: 1 84306 018 3, £19.95*

Industrial Histories

- Alldays and Onions   *ISBN: 1 84306 047 7, £24.95*
- The Life & Inventions of Richard Roberts, 1789 -1864   *ISBN: 1 84306 027 2, £29.95*
- The Textile Mill Engine   *ISBN: 1 901522 43 1, paperback, £22.50*
- Watt, James, His Life in Scotland, 1736-74   *ISBN 1 84306 045 0, £29.95*
- Wolseley, The Real, Adderley Park Works, 1901-1926   *ISBN 1 84306 052 3, £19.95*
- Morris Commercial   *ISBN: 1 84306 069 8 (Price to be announced)*

Roads & Transportantion

- Packmen, Carriers & Packhorse Roads   *ISBN: 1 84306 016 7, £19.95*
- Roads & Trackways of Wales   *ISBN: 1 84306 019 1, £22.50*
- Welsh Cattle Drovers   *ISBN: 1 84306 021 3, £22.50*
- Peakland Roads & Trackways   *ISBN: 1 901522 91 1, £19.95*

Regional/Local Histories

- Colwyn Bay, Its History across the Years   *ISBN: 1 84306 014 0, £24.95*
- Crosses of the Peak District   *ISBN 1 84306 044 2, £14.95*
- Derbyshire Country Houses: Vol 1   *ISBN: 1 84306 007 8, £19.95*
- Derbyshire Country Houses: Vol 2   *ISBN: 1 84306 041 8, £19.95*
- Historic Hallamshire   *ISBN: 1 84306 049 3, £19.95*
- Llandudno: Queen of Welsh Resorts   *ISBN 1 84306 048 5, £15.95*
- Llanrwst: the History of a Market Town   *ISBN 1 84306 070 1, £14.95*
- Lost Houses in and around Wrexham   *ISBN 1 84306 057 4, £16.95*
- Lost Houses of Derbyshire   *ISBN: 1 84306 064 7, £19.95, October 02*
- Shipwrecks of North Wales   *ISBN: 1 84306 005 1, £19.95*
- Shrovetide Football and the Ashbourne Game   *ISBN: 1 84306 063 9, £19.95*
- Well Dressing   *ISBN: 1 84306 042 6, Full colour, £19.95*